GCSE AQA
Design & Technology

Don't know your routers from your planers? Can't tell your batik from your tie-dye?
Don't worry — CGP will help you see the wood for the trees.

We've packed this book with crystal-clear study notes for the whole course,
along with brilliant diagrams to make everything super clear. We've also included
hundreds of exam-style practice questions to really test what you've taken in.

On top of all that, there's a realistic practice exam, a section on exam advice *and*
step-by-step answers... what more could you possibly want?

Complete
Revision & Practice
Everything you need to pass the exams!

Contents

Section Eight — Designing and Making

Published by CGP

Editors: Alex Billings, Christopher Lindle, Ethan Starmer-Jones.

Contributors: Ryan Ball, Sue Carrington, Juliet Gibson, Stephen Guinness, Brian Kerrush, John Nichols, Anthony Wilcock.

With thanks to Anne Ainsworth, Alex Fairer and Holly Robinson for the proofreading.
With thanks to Ana Pungartnik for the copyright research.

Photo of dressmaker on page 137: Blend Images/Shutterstock.com

ISBN: 978 1 78294 755 4

Clipart from Corel®

Printed by Elanders Ltd, Newcastle upon Tyne.

Based on the classic CGP style created by Richard Parsons.

Technology in Manufacturing

So here we go, our journey in Design and Technology <u>starts here</u> with some key ideas about <u>manufacturing</u>...

Manufacturing is a System

1) Industry involves <u>manufacturing products</u>. Manufacturing involves <u>several processes</u> that are carried out to create a <u>finished product</u> from some <u>raw materials</u> (e.g. metal, textile, paper).

2) The process of manufacturing is a <u>system</u>. Products are made as a result of a number of different <u>stages</u> taking place one after the other.

Manufacturing on a large scale often involves machinery.

3) These systems can be split up into <u>three parts</u>:

 INPUT →  **PROCESS** →  **OUTPUT**

This is all of the <u>materials</u>, <u>tools</u> and <u>equipment</u> that you start off with.

This is <u>what happens</u> to the input to change it into an output, e.g. measuring, cutting and forming.

The output is the <u>result</u> of the system — in other words, the <u>finished product</u>.

Advances in Technology can Improve Manufacturing

Technology <u>never stands still</u>. Advances can help manufacturing processes to run more <u>efficiently</u> — in recent years this has been enabled particularly by advances in <u>information and communications technology</u>.

Automation of manufacturing processes

1) Automation is the use of <u>machines</u> to do a task <u>automatically</u> without much, or any, <u>human input</u>.

2) Automation has been a <u>key development</u> in manufacturing and is present in many <u>modern factories</u>. It can be used to carry out a <u>particular process</u> or manufacture a <u>whole product</u>.

E.g. in the car industry <u>robots</u> are used to do certain processes like <u>welding</u> parts together and <u>stamping</u> out metal body panels. These robots are <u>programmed by humans</u> but then run automatically. Humans are only needed to <u>monitor</u> the robots and <u>repair</u> them if they break down.

Advantages

- Robots can <u>increase</u> the <u>speed</u> of production as they can work <u>faster</u> than humans and don't need to <u>rest</u>. This also means robots can be <u>cheaper</u> to use than human workers.
- Robots can work with <u>high accuracy</u> as they perform the same task <u>consistently</u>, whereas humans can make <u>mistakes</u> (which have costs, e.g. time and materials). Robots can therefore increase the <u>quality</u> of manufactured products and <u>reduce costs</u>.
- Robots can be used in <u>dangerous situations</u> where it would be <u>unsafe</u> for humans.

Disadvantages

- Robots can <u>replace human workers</u> so there are <u>fewer jobs</u> for people to do.
- Robots can be <u>very expensive</u> to buy.
- Robots can't carry out tasks that require <u>human judgement</u>.

3) <u>Robots</u> are continually being developed to be <u>more intelligent</u> and to allow them to work <u>more closely</u> with humans. For example, robots exist that can be <u>easily programmed</u> by humans and can '<u>learn</u>' how to do a task after being shown it by a human worker — this worker can then supervise the robot.

4) It's hoped that in the future factory robots will be developed that can deliver <u>consistent</u> and <u>accurate</u> work (as they do now) whilst also having some of the characteristics that only <u>human workers</u> currently offer (e.g. the ability to <u>make judgements</u> and come up with <u>ideas</u> for problem solving).

Smart technology in manufacturing

1) There are many examples of <u>computerised</u> machines that have been designed to carry out advanced tasks <u>accurately</u> (see CAD/CAM on page 4). <u>Smart technology</u> develops these machines further by <u>connecting them</u> with <u>other machines</u> and <u>sensors</u> in the factory so that they can <u>share data</u>.

2) These 'smart' machines can use this data to help <u>organise what tasks they need to do</u> <u>without</u> any <u>human input</u>. Smart machines can also receive <u>data</u> about <u>stock levels</u> and work out when they'll run out of materials and components (see next page).

3) Smart machines can be part of the '<u>Internet of Things</u>' — this describes the <u>connection</u> of loads of different machines, devices and other bits of technology to the <u>internet</u>. For machines this could involve <u>receiving</u> and <u>processing</u> online customer orders.

4) This use of smart technology can make automated manufacturing extremely <u>efficient</u>.

Technology in Manufacturing

Tracking of materials, tools, equipment and products

1) Factories contain <u>materials</u>, <u>tools</u>, <u>equipment</u> and the <u>products</u> themselves (both finished products and those that are partly made). Tracking of these items has been a <u>key development</u> in recent years.

2) This can be done automatically by <u>tagging</u> each box of material, tool or product etc. These tags can then be <u>scanned or detected</u> as the item moves around the factory and the status of the item kept <u>up to date</u> in a computer system.

3) Tracking the location of items has <u>many uses</u>. This includes <u>monitoring</u> the <u>stock levels</u> of raw materials and components, the <u>movement of products</u> around the factory as they're being made and the number of <u>finished products in stock</u>.

Tags like these are detected by wireless signals.

Communication systems

1) In modern automated factories, information can be passed around by <u>smart machines</u> so communication requires <u>no human input</u>.

2) <u>Human workers</u> can use <u>telephone</u>, <u>email</u>, <u>video-conferencing</u> and other internet-based communication systems to share information.

3) Workers can also use <u>devices</u> to receive information. E.g. <u>warehouse</u> workers can use a <u>device</u> (like a tablet) to <u>direct</u> them around the warehouse in the <u>most efficient route</u> to collect items in customer orders.

Specialised buildings

1) New technology means that factories are being designed to <u>incorporate smart technology</u> and also to <u>minimise</u> their <u>environmental impact</u> (e.g. using new water recycling systems to reduce water use).

2) Factories can also be made of <u>modular components</u> that can be <u>taken apart</u>, <u>moved</u> or <u>added to</u>. E.g. <u>extra production space</u> can be added if needed.

3) Technological advances can also lead to a <u>reduced need</u> for <u>room</u>. E.g. the increased use of robots and automation means that factories can take up <u>less space</u>. <u>3D printing</u> (see p.87) can use a small machine, so manufacturing can <u>move out of large factories</u> to smaller, local spaces.

Different Ways of Manufacturing can help Maximise Efficiency

Flexible Manufacturing Systems (FMS)

1) <u>Flexible manufacturing systems</u> (FMS) consist of a set of <u>different machines</u> which carry out the different stages of production. These computer-controlled <u>automated</u> systems transport materials through the different processes and eventually <u>store</u> the finished product.

2) They're called <u>flexible</u> manufacturing systems because they are <u>easy to adapt</u> — it's fairly <u>quick</u> and <u>straightforward</u> to change the system if the <u>design</u> of the product being manufactured is <u>altered</u>. E.g. the <u>layout</u> of the system can be changed and <u>new machines</u> can be added in.

3) These systems are also easy to alter if the <u>level of production changes</u>, e.g. <u>extra machines</u> can be added to the system to <u>boost production</u>.

Lean Manufacturing

1) Lean manufacturing comes from a Japanese approach to manufacturing that aims to minimise the <u>amount of resources used</u> and <u>waste</u> produced.

2) The aim of lean manufacturing is to <u>minimise costs</u> and <u>maximise efficiency</u>.

3) The <u>Just-in-Time (JIT)</u> system of stock control is a form of lean manufacturing. Materials and components are delivered <u>as they're needed</u> and used <u>as soon as they're delivered</u>.

Waste doesn't just refer to materials — it includes things like unnecessary use of time, transportation and movement of people, etc.

Advantages of JIT

- It <u>reduces</u> the amount of <u>space</u> needed for <u>storage</u> of materials and finished products (which <u>saves money</u> because you don't need to rent huge warehouses).
- It means there's less money tied up in materials that <u>aren't being used</u>.

Disadvantages of JIT

The system relies on materials and components being <u>delivered on time</u> and being <u>fault free</u> (there isn't time to return faulty goods). If these two things don't happen <u>money can be lost</u>.

These ways of manufacturing are becoming <u>more efficient all the time</u> as technology gets more advanced.

Manufacturing is becoming increasingly techie...

I know, there's a lot to learn on these first two pages. But it's all important info, so you'd best get cracking.

Production Systems — CAD/CAM

CAD and CAM are a really important part of designing products and manufacturing them. They're used in loads of different industries from food packaging to component manufacture. Now to explain what they are...

CAD is Designing Using a Computer...

1) CAD stands for computer aided design.

2) It involves designing products on a computer, rather than using a pencil and paper.

3) CAD software packages allow you to make 2D or 3D designs. Examples of CAD software include Vectorworks®, SolidWorks® and TechSoft Design®.

4) CAD helps designers model and change their designs quickly. It's easy to experiment with alternative colours and forms and you can often spot problems before making anything.

5) In 3D programs, you can view the product from all angles.

...and CAM is Making Using a Computer

1) CAM stands for computer aided manufacture.

2) It's the process of manufacturing products with the help of computers.

3) CAD software works out the coordinates of each point on the drawing. These are called x,y,z coordinates — x is the left/right position, y is forwards/backwards and z is up/down. The point where x, y and z meet is (0,0,0) — the datum.

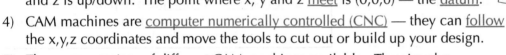

4) CAM machines are computer numerically controlled (CNC) — they can follow the x,y,z coordinates and move the tools to cut out or build up your design.

5) There are a variety of different CAM machines available. They involve the subtraction (also called wastage) or addition of material.

6) Subtraction is where material is removed from a solid block to form an object.

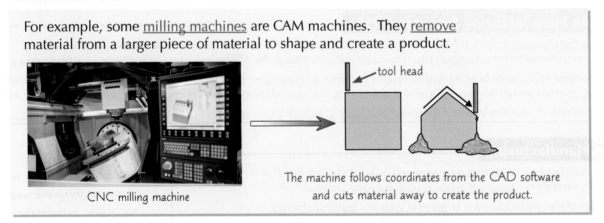

For example, some milling machines are CAM machines. They remove material from a larger piece of material to shape and create a product.

tool head

CNC milling machine

The machine follows coordinates from the CAD software and cuts material away to create the product.

7) Addition is where material is added to build up an object.

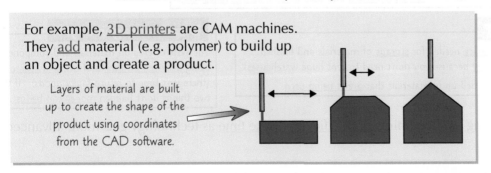

For example, 3D printers are CAM machines. They add material (e.g. polymer) to build up an object and create a product.

Layers of material are built up to create the shape of the product using coordinates from the CAD software.

3D printing is increasingly popular in industry and this looks set to continue into the future — it allows production of items closer to end users. It's also become increasingly popular with DIYers and small producers making their own custom items using 3D printers.

8) Other examples of CAM machines are CNC routers and laser cutters (see next page).

Production Systems — CAD/CAM

CAM Machines Can be Used on **Different Materials**

1) There are CAM machines out there for all kinds of jobs — fear not...

2) Some CAM machines are 2-axis — they only use x and y coordinates so can only cut out 2D shapes.

3) Others are 3-axis machines — these use x, y, and z coordinates so they can cut out 3D shapes.

> CNC routers are able to cut out either 2D or 3D shapes from a block of material using different sized cutting tools — they're either 2-axis or 3-axis machines. They can also be used to engrave things. CNC routers can be used on plastics, wood and metals.

> Laser cutters are used to cut things — they can be used on plastic, wood, cardboard, fabrics and some metals. Laser cutters on high power settings cut right through the material. On lower power settings they can be used for engraving. Laser cutters can only cut through sheet materials — they're 2 axis machines so they can't cut out 3D objects. Die cutters (see p.67) are also used to cut out shapes from sheet materials.

> 3D printers can be used for rapid prototyping — they convert your design from an image on screen into a 3D model. They can be used to print with several different materials including plastic and wax (see p.87). It's also increasingly being used to manufacture final products.

There's more about prototyping on p.147-148.

CAD/CAM is Good for **Global Companies**

1) Nowadays it's easy to communicate with people all over the world via phones and the internet — as a result businesses can operate across the world. It's also easy to transport goods around the globe.

2) These factors combined with the development of CAD/CAM has affected manufacturing in two main ways:

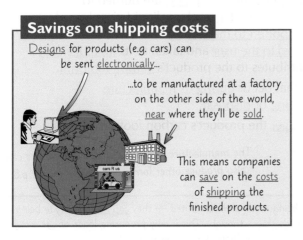

Savings on shipping costs

Designs for products (e.g. cars) can be sent electronically...

...to be manufactured at a factory on the other side of the world, near where they'll be sold.

This means companies can save on the costs of shipping the finished products.

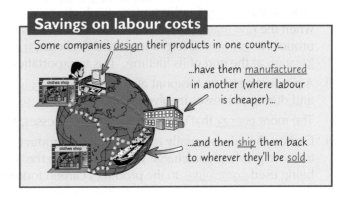

Savings on labour costs

Some companies design their products in one country...

...have them manufactured in another (where labour is cheaper)...

...and then ship them back to wherever they'll be sold.

3) CAD/CAM can offer lots of benefits to companies — these include the benefits of automation covered on page 2. However, there are drawbacks too which include the fact that computers can be affected by viruses, software problems and file corruption, which could potentially slow down production.

Remember, CAM machines make the designs that are drawn in CAD...

CAD and CAM are all the rage in designing products these days. 3D printers have been around a while but have become more widely available recently. The thought of 3D printing a custom dog bowl on your kitchen table or in your technology lab would've sounded like witchcraft not so many years ago but now it's possible.

Product Sustainability

When you make a product there's some sort of <u>impact</u> on the <u>environment</u>. Recently there's been an increasing focus on <u>minimising</u> the size of this impact to help <u>protect our planet</u>.

Sustainable Products are **Better** for the **Environment**

1) Sustainability means not causing <u>permanent damage</u> to the environment and not using up <u>finite resources</u> (ones that'll run out eventually). How sustainable a product is depends on:

> What <u>materials</u> are used to make it, for example:
> - Do they come from <u>non-finite</u> resources (ones that can be replaced) or <u>finite</u> ones?
>   E.g. most plastics are made from crude oil which is a finite resource. <u>Finite resources</u> are <u>slowly running out</u> and some could be <u>used up within 100 years</u> if new stocks aren't found.
> - Are they <u>recyclable</u> or <u>biodegradable</u> (this means they'll break down over time) — or will the used product permanently take up space in landfill?

For more about the social and ethical factors related to materials see p.44.

> The <u>processes</u> used to make the product, for example:
> - Does the process need lots of <u>energy</u>?
> - Does it create lots of <u>waste</u> or <u>pollution</u>?

<u>Wood</u> is a <u>non-finite</u> resource because trees can be <u>replanted</u>.

<u>Oil</u> is a <u>finite</u> resource — it will eventually <u>run out</u>.

2) Sustainability also depends on the <u>design</u> itself — how <u>long-lasting</u> and <u>efficient</u> the product is.

3) As you'll see on the next few pages, <u>new products</u> can have a <u>positive</u> and <u>negative</u> impact on sustainability and the environment.

All Products have a **Carbon Footprint**

1) A <u>carbon footprint</u> is the amount of <u>greenhouse gases</u> (e.g. carbon dioxide and methane) released into the atmosphere by <u>making</u>, <u>using</u> and eventually <u>reusing</u>, <u>recycling</u> or <u>disposing</u> of something at the end of its lifetime.

2) When a new product is made <u>resources</u> and the <u>finished product</u> itself are <u>transported</u>. This mileage causes the release of <u>carbon dioxide</u> because generally <u>fossil fuels</u> are burned to provide the <u>energy</u> for this transport (which emits <u>carbon dioxide</u>). E.g. carbon dioxide is released when the <u>raw materials</u> the product is made from are <u>transported to the factory</u> and when the product is <u>distributed</u> (e.g. transported to shops), <u>transported to the user</u> and taken to its <u>final disposal location</u> at the end of its lifetime. This transportation contributes to the product's <u>carbon footprint</u>.

3) A product's carbon footprint also includes the <u>emissions</u> that occur <u>during its manufacture</u> and during processes involved in its eventual <u>reuse</u>, <u>recycling</u> or <u>disposal</u>.

4) The more <u>energy</u> that's needed for these processes, the <u>bigger</u> the product's carbon footprint.

5) Carbon dioxide is usually released when a product is <u>used</u> too — e.g. when you charge your smartphone the <u>power</u> being used <u>contributes</u> to the product's carbon footprint.

The manufacture, use and disposal of products can also produce other forms of pollution — see p.8.

> So, carbon footprints aren't only linked to new products when they're being made and distributed — they <u>carry on</u> as they're being <u>used</u>. Much of a product's carbon footprint will be generated during these <u>early stages</u> though. This is why producing loads of <u>new products</u> is considered <u>bad</u> for the environment, whereas using a product for a <u>long time</u> is better for the environment.

6) The <u>bigger</u> a product's carbon footprint is, the <u>larger</u> its contribution to <u>global warming</u> (the warming of the Earth due to <u>greenhouse gases</u> in the atmosphere).

No Product Lasts **Forever** though

1) So making <u>new products</u> releases <u>emissions</u> which aren't good for the environment. The best solution would be to <u>never</u> make another <u>new product</u>. However, this obviously <u>isn't practical</u> because every product eventually comes to the <u>end of its usable life</u> — even a 'bag for life' that can be re-used many times will eventually <u>wear out</u>.

Product Sustainability

2) When a product wears out it becomes <u>waste</u> and has to be <u>disposed of</u> — this can cause <u>pollution</u>. Usually it's <u>replaced with a new one</u> too which also causes pollution... sigh.

3) Because of these things, it's important to make products using material that <u>has been</u> or <u>can be recycled</u>, or that'll <u>break down</u> when sent to landfill. For example:

   • Many materials are <u>recyclable</u>. Products often carry <u>symbols</u> to tell you if they're recyclable. <u>Paper, glass</u> and <u>most metals</u> are all <u>easy to recycle</u>. Plastics can be recycled but it can be <u>expensive</u> and they <u>can't be recycled over and over again</u> (as their structure can <u>break down</u>). For <u>safety reasons</u>, this means that some items <u>can't be made from recycled plastics</u>, e.g. medical supplies.

   • Some materials are <u>biodegradable</u> — they will <u>rot away</u> naturally (in a compost heap, say). <u>Wood</u> and <u>paper</u> are biodegradable and so are <u>some new plastics</u> (see next page).

4) Products made from <u>lots of components</u> and several <u>different materials</u> are awkward to recycle, because it's difficult to separate all the parts before you can recycle them.

5) However, '<u>Design for Disassembly</u>' tries to help with this problem. This is when a new product is designed so it can be <u>easily taken apart</u> at the <u>end</u> of its <u>lifetime</u> — this allows the <u>parts and materials</u> to be <u>reused or recycled</u> to make new products.

6) There's a <u>big drive at the moment</u> for all electronic and electrical goods to be made this way. It's an example of new products being designed to have a <u>reduced environmental impact</u> — there's more about this on the next page.

Some Products **Aren't** Designed to **Last** but Some Are

1) Some products are designed to <u>become obsolete</u> (<u>useless</u>) quickly. E.g. a disposable razor becomes blunt after a few uses and its blade can't be sharpened or changed. This is <u>planned obsolescence</u>.

2) Products with <u>up-to-the-minute</u> designs or technology become obsolete quickly because they <u>go out of fashion</u>. E.g. people often replace their <u>mobile phone</u> when a new, fancier model comes out.

3) Built-in obsolescence is generally <u>bad</u> for the environment — because more <u>materials</u> and <u>energy</u> are used to make <u>replacement</u> products.

4) However, products <u>can</u> be <u>designed to last</u> and have less of an impact on the environment. This involves making the product <u>durable</u>, and designing it so that parts can be <u>maintained</u> and <u>repaired</u> or <u>replaced</u>. This is known as <u>design for maintenance</u>.

   • Most <u>household appliances</u> are designed to be <u>maintained</u> and <u>repaired</u> — this makes sense as they're <u>expensive</u> and just replacing them when they break would be <u>wasteful</u> as they are made of lots of parts.

   • For example, washing machines can be maintained (e.g. cleaning filters) and they can be <u>repaired</u> by trained technicians when they break, e.g. by installing a <u>new part</u> rather than having to replace the whole thing.

   • Another example is the idea of making <u>modular electronics</u>. This has been proposed for <u>mobile phones</u> and other electronics that are often updated like <u>tablets</u>, <u>laptops</u>, <u>cameras</u> and <u>music players</u>.

   • The idea is that electronics are made up of different parts (or modules) that are designed so that they can be <u>individually upgraded and replaced</u>. E.g. with a modular phone you could <u>replace</u> the <u>processor</u> for a <u>faster one</u> when it starts to get a bit slow or the <u>battery</u> when it wears out. This is an alternative to getting a <u>completely new phone</u> when only one part of it breaks down, which <u>cuts down</u> a lot of <u>electronic waste</u> (which often ends up going to landfill because it's difficult to recycle).

Design for DISassembly helps with the DISposal of a product...

...as it allows products to be taken apart easily so that parts can be recycled or re-used. Don't confuse this with design for maintenance, which involves making products durable and repairable.

Product Sustainability and Social Issues

Just a bit more on the <u>environmental impacts</u> of new products and then onto the <u>social impacts</u>.

Continuous Improvement of Products can be Bad and Good

1) Continuous improvement is a process by which manufacturers are constantly trying to <u>improve their products</u>, e.g. by changing the design of a product so it <u>incorporates new technology</u> (see page 10).

2) The continuous improvement of products can be <u>environmentally damaging</u> for many reasons, including:

- It can encourage consumers to <u>replace existing products</u> with new ones, which can lead to older models being <u>disposed of</u>. New, replacement products that are made have a <u>carbon footprint</u> (see p.6).
- New products being manufactured, packaged, transported and eventually disposed of can result in the <u>increased usage of finite resources</u> and <u>environmental damage</u> linked to the collection of these resources (see p.58).

3) However, manufacturers can also use the process of continuous improvement to make changes to their products that mean they have a <u>reduced environmental impact</u>. For example, new electronic products can be made using <u>more efficient</u> components than older models, therefore having <u>lower carbon footprints</u> (meaning they contribute less to global warming).

Many appliances have energy efficiency <u>ratings</u>, e.g. an A-rated fridge is more efficient than a D-rated one. Generally <u>newer products</u> have <u>more efficient</u> ratings than older ones and this can represent <u>big power savings</u> when they're being used.

4) Improvements that have a <u>positive impact</u> on the environment can also relate to <u>other aspects</u> of products and their manufacture, including:

- <u>Improving</u> the <u>energy efficiency</u> of <u>manufacturing processes</u> used to make products.
- <u>Avoiding</u> use of <u>environmentally damaging chemicals</u> (e.g. CFCs which contribute to global warming).
- <u>Minimising</u> the <u>amount of material</u> used in products where possible. This means <u>fewer resources</u> are consumed and products are <u>lighter</u> (so less energy is needed to transport them).
- <u>Avoiding</u> use of <u>materials</u> that are produced from <u>unsustainable sources</u> — e.g. timber sourced from rainforests.
- Using <u>more renewable</u> and/or <u>biodegradable</u> materials — e.g. bioplastics can be made that are made of (renewable) plant materials that also biodegrade (rot away) fully.
- Making products from materials that can be <u>recycled</u> and that are <u>designed for disassembly</u> (see previous page). This means that <u>fewer new resources</u> are needed, and often <u>less energy</u> is used — e.g. recycling old food cans takes <u>much less</u> energy than mining and processing <u>new metal</u>.
- Making products that are <u>designed to last</u> (see previous page).

You can Carry Out a Life Cycle Assessment

You can work out the potential <u>environmental impact</u> of a product by doing a <u>life cycle assessment (LCA)</u>. These look at each <u>stage</u> of the <u>life</u> of a product — from the raw materials to when it's disposed of.

Choice of material

1) <u>Hardwoods</u> are often obtained from natural <u>rainforests</u>. Felling the trees destroys the habitat of pretty much everything living there (including people). <u>Softwoods</u> are a <u>greener choice</u>. They're usually from <u>managed plantations</u> — so more trees are planted and grow quickly to replace them. <u>Recycled</u> wood is also a good choice for the environment.

2) <u>Metals</u> have to be <u>mined</u> and <u>extracted</u> from their ores. Most <u>plastics</u> are made using <u>crude oil</u>, which is a <u>finite resource</u> so <u>can't be replaced</u>. These processes need a lot of <u>energy</u> and cause a lot of <u>pollution</u>.

Some products can be <u>recycled</u> — the materials can be used again in new products.

Manufacture

1) <u>Manufacturing</u> products uses a lot of <u>energy</u> and other resources. It can also cause a lot of <u>pollution</u> (e.g. some processes produce toxic gases).

2) You also need to think about <u>waste</u> material and how to <u>dispose</u> of it.

Product disposal

1) Products are often <u>disposed</u> of in a <u>landfill</u> site at the end of their life.

2) This takes up space and <u>pollutes</u> land and water (e.g. when paint washes off a product and gets into rivers).

Using the product

<u>Using</u> the product can also damage the environment. E.g. <u>electrical products</u> use electricity generated by burning <u>fossil fuels</u>, and <u>paint</u> can give off <u>toxic fumes</u>.

Product Sustainability and Social Issues

You Need to Know the 6 Rs

You can use the 6 Rs when designing to help reduce the impact that new products have on the environment and make the whole process more sustainable. There are things for consumers to think about too.

1) REPAIR
It's better to fix things instead of throwing them away. E.g. repairing a mobile phone can sometimes cost the same as buying a new one, but is better for the environment. Manufacturers can still make a profit by selling replacement parts.

2) RE-USE
Customers can extend a product's life by passing it on or using it again. Some people reuse products for other purposes, e.g. using an old car tyre to make a swing.

3) RECYCLE
Recycling uses less energy than obtaining new materials, e.g. by extracting metal. Products made from more than one material should ideally be easy to separate into recyclable stuff — clear 'recycle' labelling helps with this.

4) RETHINK
You should think about your design carefully — you might be able to make the product in a different way, e.g. a radio that you wind up instead of running off batteries.

5) REDUCE
Making long-lasting, durable products like rechargeable batteries reduces the number of products customers need to buy. It also means that manufacturers can cut down on energy use and transport.

6) REFUSE
You can refuse to buy a product if you think it's wasteful — e.g. if it has lots of unnecessary packaging, has travelled long distances round the world to the UK, or will be inefficient or costly to run.

Products have a Social Footprint

As well as having an environmental impact, the design and manufacture of products also impacts people.

Working conditions
- Firms have a moral and legal responsibility to provide safe working conditions for their employees.
- This is particularly relevant to people working in dangerous situations (e.g. agriculture and manufacturing). Keeping workers safe in these jobs can involve providing them with a safe working environment (e.g. factories designed to be as safe as possible), sufficient protective equipment (e.g. safety goggles) and training about how to work safely.
- There are initiatives for better working conditions. Companies that sign up to the Ethical Trading Initiative (ETI) agree to meet certain standards for working conditions, right across the supply chain — from the extraction of raw materials, through product manufacture, to the distribution of the products to the consumer.

Health impacts
- As you've seen on the past few pages, making products can release pollution (e.g. harmful chemicals). These can negatively affect the oceans and the atmosphere. For example, dioxins are made when fibres are bleached during the manufacture of paper, card and textiles. These chemicals are highly toxic to both animals and humans.
- Reducing levels of pollution is important because of the potential danger to human health that it can have. Efforts are underway to clean up manufacturing processes to reduce (or stop completely) the release of harmful pollutants in the environment. These include legal controls of certain chemicals and tough restrictions on what factories can release into the environment.

The 6 Rs can help to reduce a product's environmental impact...

...but only if the designer and consumer follow them. Try testing your knowledge of the 6 Rs. Cover up the page, scribble them down, and describe each of them. If you can't remember all 6, revise them again and have another go — keep going until you get them all scribbled down.

Products in Society

New technology can impact <u>enterprise</u>, <u>people</u>, <u>culture</u> and <u>society</u> as a whole. <u>Read on</u> to find out how...

Entrepreneurs Take Advantage of Business Opportunities

1) <u>Enterprise</u> involves identifying new business opportunities, and then taking advantage of them. There's always a <u>risk of failure</u>, but the <u>reward</u> for a successful enterprise activity is <u>profit</u>.

2) Enterprise can involve <u>starting up</u> a <u>new business</u>, or helping an <u>existing one</u> to <u>expand</u> by coming up with new ideas (e.g. a new product). An <u>entrepreneur</u> is someone that does this.

3) When an entrepreneur or a business come up with something new this is called <u>innovation</u>. There are some <u>innovations</u> in business you need to know about:

Crowdfunding

- To start a new business or just launch an idea entrepreneurs need to <u>raise money</u>. Traditionally you could do this by applying for a <u>loan from a bank</u> or finding an <u>investor</u> (think Dragon's Den), but technology has provided another way called <u>crowdfunding</u>.
- Crowdfunding involves using websites (e.g. Kickstarter) to promote an idea to a huge number of people. People can then <u>choose to invest</u> an amount of money to help fund the idea — these people are called <u>backers</u>. Backers often get a <u>reward</u> for their investment — this could be a <u>free gift</u>, a <u>discount</u> on the product being funded or perhaps a <u>percentage share</u> of the new company once it has been funded.

Virtual marketing and virtual retail

- <u>Marketing</u> is the promotion of a product or service (e.g. magazine adverts). <u>Retail</u> is selling products and services to customers (e.g. selling products in high street shops). On the internet it's possible to do both of these things <u>virtually</u>.
- <u>Virtual marketing</u> includes promoting a product/service on social media (e.g. Facebook®), via email, and moving a product/ service nearer to the top of the page in search engine results.
- <u>Virtual retail</u> is all about selling products and services on the internet. Many businesses sell their products <u>online</u> already but in the future it may become a <u>more realistic</u> experience with <u>virtual reality shops</u> that you can move around in from the comfort of your own living room...

Co-operatives and Fairtrade

- A co-operative is a type of business that is <u>owned and run</u> by its <u>members</u>. The members make the decisions about how the business is run and <u>profits are shared out</u> between them. Co-operatives include <u>small enterprises</u> like local shops and <u>large businesses</u> (e.g. The Co-operative Group).
- Groups of <u>farmers</u> often form co-operatives. This is when farmers work together so that they have stronger <u>negotiating</u> power with trading partners and can get <u>higher prices</u> for their produce. Co-operatives are a feature of <u>Fairtrade</u> which aims to ensure farmers are paid a fair price — see p.132.

Manufacturers Try to Continually Improve Products

1) Established product manufacturers are always looking for ways to make <u>more money</u>. One way to do this is by <u>improving</u> how they make their products, e.g. improving the <u>design</u> so that the product can be made more <u>easily</u>.

2) Manufacturers also redesign products in response to <u>market pull</u> or advances in <u>technology</u>...

Market Pull is About What Consumers Want

1) Designers often <u>design stuff</u> (and manufacturers make it) to satisfy the <u>wants</u> and <u>needs</u> of consumers — <u>consumer demand</u>.

2) Changing <u>fashions</u> and <u>social attitudes</u> affect the kind of products people want — consumer demand <u>won't</u> always be for the same <u>things</u> or <u>styles</u>.

E.g. the car was invented to transport people from A to B, but now some consumers expect it to be more of a <u>status symbol</u>, demanding <u>luxury extras</u> like climate control, reversing cameras and fancy stereos.

Technology Push is About What Manufacturers Can Provide

1) In industry, research and development departments are always coming up with <u>new technologies</u>, <u>materials</u> and <u>manufacturing techniques</u>. This can drive the design of new products or <u>improve existing ones</u>.

2) Using new technology might make an existing product <u>cheaper</u>, <u>better</u> at its <u>function</u>, <u>more efficient</u> or <u>nicer-looking</u> — all things which will make products <u>more desirable</u>.

This process can have environmental impacts as it promotes increased consumption and waste — see p.8.

E.g. computers started off as <u>huge mechanical 'adding machines'</u>. Now, thanks to technologies including the microchip, they're <u>small</u> but really <u>fast and powerful</u>.

Products in Society

Technology Affects what Jobs People Do

1) Technology is always developing and this has an impact on the <u>jobs that are available</u> to do.

2) Increasing <u>automation</u> and the <u>use of machines</u> in manufacturing has <u>reduced</u> the number of people that are needed to carry out some <u>manual tasks</u> — this was covered earlier in the section (see p.2).

3) However, developments in technology have also opened up <u>new job opportunities</u>. For example, the development of <u>computers</u> and the <u>internet</u> has created lots of jobs. Workers are needed to <u>design</u> computers, <u>write programs</u> for them to run, design <u>computer games</u>, maintain <u>websites</u>, <u>teach</u> people how to use computers and devices... the list goes <u>on and on</u>.

4) Technology has also led to jobs <u>changing</u> as a result of the increased use of technology in the workplace.

- Many <u>delivery drivers</u> use a device with <u>satellite navigation</u> to work out their <u>most efficient route</u> to complete their deliveries. This means they don't have to use a <u>map to navigate</u> or spend time working out the <u>best route</u> to do their deliveries.
- Workers on manufacturing lines may <u>supervise a robot</u> rather than doing a technical task themselves.

Technology can Also Have an Impact on Culture

1) The <u>culture</u> of a particular <u>country</u> or <u>group of people</u> covers everything from their <u>religion</u>, <u>beliefs</u> and <u>laws</u> to their <u>language</u>, <u>food</u>, <u>dress</u>, <u>art</u> and <u>traditions</u>. Phew.

2) If you're designing a product aimed at a <u>specific</u> target market, you'll have to take into account the <u>views and feelings</u> of people from that particular culture — that way you don't <u>isolate or offend</u> anybody for <u>political</u>, <u>religious</u>, <u>gender</u> or <u>cultural</u> reasons.

3) New technology has the potential to <u>offend certain people</u>. For example, if a new material is developed that is made using <u>animal products</u> (e.g. animal fats), vegans and vegetarians may <u>choose not to buy</u> products made of the material.

4) New technology can also impact <u>fashion</u> and <u>trends</u>.

- Fashion itself is continually affected by <u>new materials and techniques</u>. For example, in the 1960s when stretchy, body-hugging LYCRA® was introduced it revolutionised sportswear. New textiles and ways of making clothes will continue to give <u>new possibilities</u> for fashion.
- Technology can also have an impact on <u>fashion trends</u>. The <u>internet</u> allows people to find out about fashion trends that are happening all over the world and new clothes can be seen by a <u>global audience</u> of people, e.g. via social media websites and blogs.

Products can Impact Particular Groups in Society

1) Products that are made can be <u>designed</u> to avoid having a <u>negative impact</u> on other people by being <u>sensitive to their needs</u>. Groups of people to consider include <u>disabled</u> and <u>elderly</u> people (see p.126).

2) Technology designed to suit <u>disabled and elderly users</u> includes <u>more accessible mobile phones</u> and <u>apps</u>. For example, mobile phones are sold with <u>large</u>, <u>easy-to-read buttons</u> that are easier to use for people with <u>less mobile hands</u> and/or <u>poor sight</u>. Apps that enable devices to be controlled entirely by <u>voice commands</u> have been developed to assist people with <u>physical disabilities</u>.

3) When designing products, people of different <u>religious groups</u> should also be considered.

4) If people believe that <u>religious symbols</u> have been misused or abused on a product this is likely to cause offence. For example, if symbols from a religion that <u>doesn't permit</u> alcohol consumption were used on the packaging of an <u>alcoholic drink</u> this is likely to be offensive.

New technology has an impact on all sorts of things...

...from business to culture and loads more. Make sure you know your market pull from technology push.

Powering Systems

Energy is needed to manufacture products and to power systems. There are many ways of providing energy. You need to know about how they work, and the arguments for and against using different energy resources.

Fossil Fuels are Finite (Non-Renewable)

1) Fossil fuels are natural resources that form underground over millions of years. There are three main types — coal, oil and gas.

2) They're typically burnt to provide energy. For example, power stations burn fossil fuels to produce heat. This heats water to create steam, which drives a turbine. This turns a generator, generating electricity.

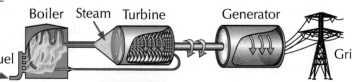

For

1) Fossil fuels are reliable (at the moment) — there's enough to meet current demand.

2) Running costs of power plants and extraction costs for fossil fuels are quite low — so they're a fairly cost-effective way to produce energy.

Against

1) They're finite — they'll eventually run out.

2) The extraction of fossil fuels has social and environmental impacts (see p.58).

3) They release greenhouse gases when they are burned, which causes lots of environmental problems (see p.6).

Nuclear Power Produces Dangerous Waste

Nuclear power stations work in a similar way to fossil fuel power stations. The difference is that a process called nuclear fission produces the heat to create the steam that drives the turbines.

For

1) Nuclear power is reliable.

2) Nuclear fuel is quite cheap.

3) Nuclear power is clean — it produces very low levels of greenhouse gases.

Against

1) It uses fuel sources that are finite, e.g. uranium.

2) Nuclear power plants cost a lot to build, maintain, and decommission (shut down and make safe).

3) The waste produced is dangerous and difficult to dispose of.

4) Nuclear power always carries the risk of a major catastrophe, e.g. the Fukushima disaster in Japan.

Non-Finite (Renewable) Energy is Better for the Environment

Non-finite energy resources provide energy that won't run out — it can be renewed as it is used:

1) Wind power involves putting turbines in exposed places where they can take advantage of the wind. Each turbine has a generator inside it — the rotating blades turn the generator and produce electricity.

2) Solar cells generate electric currents directly from sunlight.

3) Tidal barrages are big dams built across river estuaries, with turbines in them. The flow of the tide spins the turbines and generates electricity.

4) Hydro-electrical power uses a dam to flood a valley. Water trapped in the top of the valley (the higher reservoir) is only allowed into the lower reservoir through turbines connected to generators.

5) Biomass involves burning waste wood or crop material to turn water into steam, which turns turbines, etc.

For

1) Renewable energy resources generally have a smaller environmental impact than other energy resources — once they've been set up, they provide clean energy.

2) After the initial set up costs, the energy provided is usually free.

Against

1) Initial set up costs are often quite high. They also generally don't produce as much energy as finite sources, so the time taken to pay back the initial expense is quite long.

2) Some types rely on external factors like the sun and wind, so they're less reliable.

3) Some people think renewable energy resources (e.g. wind turbines) spoil the landscape and look ugly. Wind farms can also be noisy, which can affect people living locally.

4) Some types of renewable energy resources can have a big impact on the environment, e.g. valleys are flooded for hydro-electric power, which can cause possible habitat loss.

Powering Systems

There is a **Drive** to use **More Renewable Energy**

Most of our electricity is produced using fossil fuels and nuclear power. Recently, there has been a move towards using more renewable energy resources. This has been triggered by many things...

- We now know that extracting and burning fossil fuels damages the environment.
- Many people think it's better to learn to get by without non-renewables before they run out.
- Improved efficiency in renewable power production mean it's becoming a more attractive option.
- Pressure from other countries and the public has meant that governments have begun to introduce targets for using more renewable resources, and for cutting down on carbon dioxide emissions (see p.6).

However, many people are opposed to the idea of renewables replacing fossil fuels, due to factors such as initial cost, reliability, the effect on the landscape, etc.

Energy Often Needs to be **Stored**

1) Most large power stations have to be kept running all night even though demand for electricity is very low. This means there's extra electricity being generated at night.

2) A kinetic pumped storage system is one of the best ways to store this extra energy for when it is needed during periods of peak demand.

3) The system operates using a hydro-electric power station. Here's how:

- Rather than wasting it, 'spare' night-time electricity generated by other power stations is used to pump water in the lower hydro-electric reservoir up to the higher reservoir.
- This water can then be allowed to flow through the turbines during periods of peak demand to generate electricity — this supplements the steady delivery from the other power stations.

Pumped storage uses the same idea as hydro-electric power but it isn't a way of generating power — it stores the energy that has already been produced by other power stations.

Batteries are Another Way of **Storing Energy**

1) Batteries store chemical energy — chemical reactions are used to generate an electric voltage, which pushes a current through a circuit.

2) They're often used to power portable electronics.

Electric currents, voltages and circuits are covered in more detail on p.30.

Alkaline batteries

1) Alkaline batteries are normally disposable — once they've gone flat, they can't be used again.

2) They're recyclable, last a long time, and leak less than some disposable batteries though, which helps to reduce their environmental impact.

3) Alkaline batteries have a power output that gradually decreases over time.

4) They're typically used for toys, remote controls, torches and clocks.

Rechargeable batteries

1) A rechargeable battery can be charged when it goes flat — electricity is used to reverse the chemical reaction that occurs whilst the battery is in use. This reforms the chemicals that are needed for the battery to work again.

2) They're more expensive than alkaline batteries, but they can be used again and again — this makes them cheaper in the long run.

3) They are more environmentally-friendly than disposable batteries because they can be used more than once.

4) Their power output remains constant until they run flat.

5) They're built in to some products (e.g. mobile phones, laptops, electric cars).

Storing energy means it can be used when it's needed most...

Make sure you understand the two main ways of storing energy — batteries and kinetic pump storage systems.

Warm-Up and Worked Exam Questions

Congratulations, that's it for this section. The only thing standing between you and a tea break is this series of question pages to make sure that you're an expert in this bit of Design and Technology.

Warm-Up Questions

1) Give two advantages of using a Just-In-Time (JIT) system.
2) What is the link between the size of a product's carbon footprint and its impact on global warming?
3) What is meant by planned obsolescence?
4) Name the 6 Rs.
5) Manufacturers are always looking for ways to make their products better. For example, they often change their designs to incorporate new technology. What is this process called?
6) Describe the difference between renewable and non-renewable energy resources.

Worked Exam Questions

1 Products that are designed to be maintained can have less of an impact on the environment than products which haven't. Explain what is meant by 'design for maintenance'.

It means making a new product durable and designing it so that parts can be maintained and repaired or replaced.

[2 marks]

2 A CNC router is an example of a CAM machine. They can be either 2-axis or 3-axis machines.

a) Give **two** ways that a CNC router can be used.

1. *It can be used to cut out shapes from a solid block of material.*

2. *It can be used to engrave objects.*

[2 marks]

b) Explain the difference between 2-axis and 3-axis CAM machines.

2-axis machines only use x and y coordinates so they can only cut out 2D shapes, whereas 3-axis machines use x, y and z coordinates so they can cut out 3D shapes.

[2 marks]

3 State **two** arguments against the use of fossil fuels as an energy resource.

1. *They're finite so will eventually run out.*

You could have also written about the negative environmental and social impacts caused by the extraction of fossil fuels.

2. *They release greenhouse gases when they are burned, which contributes to environmental problems such as global warming.*

[2 marks]

Exam Questions

1   Which **one** of the following is a feature of alkaline batteries?

   A   They can be used more than once. ☐
   B   They are built in to products such as mobile phones. ☐
   C   Their power output gradually decreases over time. ☑
   D   They are expensive. ☐

   [1 mark]

2   Which **one** of the following is a process of designing products using a computer?

   A   CAM ☐
   B   CNC ☐
   C   CAD ☑
   D   3D printing ☐

   [1 mark]

3   Which **one** of the following does the process of crowdfunding **not** involve?

   A   People called backers that invest in an idea. ☐
   B   An application to a bank for a loan. ☑ ?
   C   Promoting a business idea to try and attract investment. ☐
   D   A website to promote a business idea to potential investors. ☐

   [1 mark]

4   A power tool company regularly releases new, improved versions of the tools that it manufactures.

   a) State **two** reasons why the company regularly releasing new versions of their tools can be environmentally damaging.

   1. Any unsold versions of the old tools have to be thrown away

   2. Increased rates of fossil fuels during production.

   [2 marks]

   b) Explain why newer power tools can have a reduced environmental impact compared to older power tools.

   The newer power tools were made to be more efficient which means it has less of a carbon footprint.

   [2 marks]

Section One — Key Ideas in Design and Technology

Exam Questions

5 Robots are used in many modern factories.

a) State **two** reasons why a company might choose to use robots rather than humans in manufacturing.

1. ....

....

2. ....

....

[2 marks]

b) Give **one** reason why robots aren't always able to replace human workers in manufacturing.

....

....

[1 mark]

6 **Figure 1** shows some paper envelopes and some plastic bubble wrap.

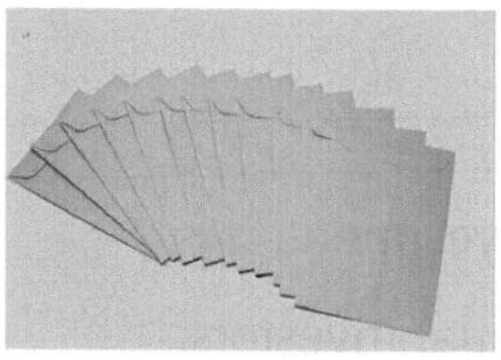

envelopes    bubble wrap

Figure 1

a) The products shown in **Figure 1** are recyclable.
Give **one** reason why this is good for the environment.

....

....

[1 mark]

b) The envelopes shown in **Figure 1** are biodegradable but the bubble wrap is not.
Why does this make the envelopes more sustainable than the bubble wrap?

....

....

[2 marks]

c) A padded envelope is a paper envelope with a layer of bubble wrap inside.
What makes padded envelopes difficult to recycle?

....

....

[2 marks]

Revision Questions for Section One

Well, that's <u>Key Ideas in Design and Technology</u> all wrapped up — time to see <u>how much</u> you can remember.
- Try these questions and <u>tick off each one</u> when you <u>get it right</u>.
- When you've done <u>all the questions</u> for a topic and are <u>completely happy</u> with it, tick off the topic.

Technology in Manufacturing (p.2-3) ☑

1) a) Explain what is meant by a Just-in-Time system.
   b) State one disadvantage of using a JIT system.
2) Describe what is meant by a flexible manufacturing system.
3) Describe one way in which smart technology can be used in manufacturing.

Production Systems — CAD/CAM (p.4-5) ☑

4) What do CAD and CAM stand for?
5) What is meant by machines being 'computer numerically controlled'?
6) Give an example of a CNC machine.
7) Describe the difference between CAM machines that carry out subtraction and CAM machines that carry out addition of material.
8) Give an example of a material that a 3D printer can print with.
9) Describe how using CAD/CAM can save on shipping costs for a business.

Product Sustainability and Social Issues (p.6-9) ☑

10) a) Define the term 'sustainability'.
    b) State two factors that influence the sustainability of a product.
11) Describe what the carbon footprint of a product is.
12) What is meant by 'Design for Disassembly'?
13) What is the purpose of a life cycle assessment?
14) Which of the following is not one of the 6 Rs?
    A: Repair   B: Reassess   C: Rethink   D: Recycle
15) Give two ways that firms can help keep their employees safe.

Products in Society (p.10-11) ☑

16) Briefly describe what is meant by the following terms:
    a) Co-operative   b) Virtual retail
17) Tim is designing a mobile phone. Give two ways how technology push could affect his design.
18) Why should people's views be considered when products are designed for a particular culture?

Powering Systems (p.12-13) ☑

19) List one advantage and one disadvantage of nuclear power.
20) Briefly describe the main steps involved in generating electricity from a fossil fuel power station.
21) A company is considering ways to reduce its energy bills. It is considering building either a single wind turbine nearby, or installing solar panels on top of their main factory.
    a) Suggest two reasons why residents living near the turbine may prefer the use of solar power.
    b) Suggest one reason why the company may choose a wind turbine over solar panels.

Properties of Materials

The <u>properties</u> of <u>materials</u> are their <u>characteristics</u> — all materials can be <u>described</u> in terms of their properties.

Different **Materials** Have Different **Properties**

1) The <u>properties</u> of a material determine what it's <u>useful for</u>. Thinking about these properties is super important when it comes to <u>designing</u> a <u>new product</u>.

2) Materials have <u>working properties</u>. Here are a few you'll need to <u>know</u> for the exam...

Strength
Strength is the ability to <u>withstand forces</u> without <u>breaking</u>. For example:
1) The rope in a tug-of-war resists <u>pulling</u> forces.
2) Bridge supports resist <u>compression</u> forces.
3) A surfboard resists forces trying to <u>bend</u> it.
4) <u>Fabrics</u> that contain <u>Kevlar</u>® fibres (see p.39) are really strong and <u>resistant to abrasion</u> — so they're used in <u>motorcycle clothing</u>.

These motorcycle gloves contain Kevlar®.

Hardness
1) This is the ability to withstand <u>scratching</u>, <u>abrasion</u> or <u>denting</u>.
2) It's very important for tools that cut, like <u>files</u> and <u>drills</u>.

Toughness
1) If a material is <u>tough</u>, it is hard to <u>break</u> or <u>snap</u> — the material changes shape a bit instead.
2) <u>Armour</u> and <u>bulletproof vests</u> need to be tough.

Elasticity
1) Elastic materials can <u>stretch and bend</u> and <u>return</u> to their <u>original shape</u>.
2) A <u>spring</u> has good elasticity.

Malleability
1) Materials that are malleable can be <u>bent</u> and <u>shaped</u>.
2) Most <u>metals</u> are malleable — they can be hammered into <u>thin sheets</u> without breaking.

Ductility
<u>Ductile</u> materials can be <u>drawn</u> into a <u>wire</u>.

3) Materials also have <u>physical properties</u>. For example:

Electrical conductivity
1) Electrical <u>conductors</u> let electricity <u>travel through them</u> easily. Electrical <u>insulators</u> don't.
2) <u>Electrical wires</u> need to be <u>conductors</u>, but the <u>coating</u> around the wires must be <u>insulating</u>.
3) <u>Metals</u> are good electrical <u>conductors</u>. <u>Plastics</u> tend to be good <u>insulators</u>.
4) Some <u>fabrics</u> are <u>blended</u> or <u>coated</u> with an electrically conductive material (e.g. a <u>metal</u>) — <u>gloves</u> can have <u>electrically conductive fingertips</u> so you can still operate <u>touchscreens</u> with them on.

Copper is commonly used in electrical wires — it is both ductile and very electrically conductive.

Thermal conductivity
1) Thermal <u>conductors</u> let <u>heat travel</u> through them <u>easily</u>. Thermal <u>insulators</u> don't.
2) <u>Metals</u> are good thermal <u>conductors</u>. <u>Plastic</u>, <u>board</u> and <u>wood</u> are good thermal <u>insulators</u>.
3) <u>Pans</u> must be made from good thermal <u>conductors</u>, but their <u>handles</u> are often made from thermal <u>insulators</u>.

Properties of Materials

Fusibility

1) Materials with a high fusibility have low melting points — only a small amount of heating is required to convert these materials to liquids.

2) For example, solder has a high fusibility — this allows it to melt before the metals that are being soldered together (which have a lower fusibility).

Density

1) The density of a material is a measure of its mass per unit volume.

2) A table made of solid metal would likely be heavier to carry than an identical table made from plastic. This is because metals tend to be denser.

3) Density often has units of kg/m³.

Absorbency

1) Fibres and fabrics that are absorbent are good at soaking up moisture. Paper towels are a good example.

2) Absorbent materials can be dyed easily, but they also dry slower and are vulnerable to stains.

3) Natural fibres (e.g. wool, cotton and cellulose fibres that make up paper) are absorbent.

4) Synthetic fibres (e.g. polyester and LYCRA®) are not absorbent.

You can also describe how resistant to moisture a material is. Materials that are poor at absorbing have a high resistance to moisture.

Non-Metals and Metals Have Different Physical Properties

Non-Metals

Non-metals tend to have a very different set of physical properties to metals.
Compared to metals, they are generally:

- More brittle (the opposite of tough)
- Not always solid at room temperature
- Poor electrical conductors
- Dull looking
- Less dense

Non-metals include all materials that don't contain metal — e.g. plastics, paper, textiles and wood.

Metals

Metals tend to have similar basic physical properties to one another. They're usually:

- Strong
- Malleable
- Good conductors of heat and electricity
- Not very fusible (have high melting and boiling points)

Metals and non-metals also have a wide range of chemical properties.

An Alloy is a Mixture

1) Alloys are a mixture of two or more metals, or a metal mixed with one or more elements.

2) The alloy is a new material — it has different properties to the individual metals it's made of.

3) Alloys are developed to have a specific set of properties. This is often done with future products in mind.

The properties of a material determine what it can be used for...

It's really important that you know the differences between each of the properties listed here — it'll crop up a lot in this section as we'll be covering the properties of different types of material. Have a go at drawing a table, listing each of the properties in one column. In the next column, describe what each property means, name a product/material that has this property, and explain why it's useful. Remember to check your work, and repeat for any of the bits you've got wrong.

Paper, Board and Timber

Time to move on to materials made from <u>plants</u> and <u>trees</u> — that's right, it's <u>paper</u>, <u>board</u> and <u>timber</u>.

There are **Various** Types of **Paper** to Choose From...

1) Paper is pretty useful for <u>writing</u> and <u>sketching</u> (no, really).

2) There are lots of sorts of <u>paper</u> — each designed for a particular use.

3) <u>Cartridge paper</u> is high quality and has a <u>textured</u> surface — it's great for sketching with different drawing materials like pencils, crayons and inks.

4) <u>Layout paper</u> is <u>thin</u> and <u>translucent</u> (you can see light through it) and is used for <u>general design work</u> — particularly sketching ideas.

5) <u>Tracing paper</u> is <u>semi-transparent</u>, and is used to <u>copy images</u>.

6) <u>Grid paper</u> may have a <u>square</u> or <u>isometric pattern</u> printed on it. Square grid paper's useful for <u>orthographic and scale drawings</u> (see p.141-142) and isometric paper's good for <u>isometric drawings</u> (see p.139).

7) <u>Bleed-proof paper</u> is used by designers when drawing with <u>felt-tips</u> and <u>marker pens</u>. The ink doesn't spread out (<u>bleed</u>) — it stays put.

Square grid paper

Isometric grid paper — this paper also comes without lines (i.e. just dots where the lines cross).

Ink bleeds on some paper because the paper fibres suck the ink away.

...and a Selection of **Boards** too

1) The weight of paper and board is measured in <u>gsm</u> (grams per square metre). Above <u>200 gsm</u>, it's not paper any more — it's <u>board</u> (also known as <u>card</u> or <u>cardboard</u>).

2) Board is often used in packaging because of its <u>low cost</u> compared to other packaging materials, and its <u>high strength-to-weight ratio</u>.

<u>Solid white board</u> has a high quality bleached surface, which is ideal for <u>printing</u>. It's used loads for <u>primary packaging</u> — packaging for individual items. (<u>Secondary packaging</u> is used to contain lots of the same item — see below.)

<u>Ink jet card</u> is card used for <u>ink jet printing</u>. It's designed so that the ink <u>doesn't bleed</u> — this allows the printed image to be <u>sharply defined</u> and of a <u>high quality</u>.

<u>Corrugated card</u> is made up of a <u>fluted inner core</u> sandwiched between <u>two outer layers</u> (the <u>liner</u>), which can be <u>printed</u> on. The flutes add <u>strength</u> and <u>rigidity</u> — this is useful in a lot of <u>secondary</u> packaging to <u>protect</u> products during transit.

<u>Duplex board</u> has a <u>different colour</u> and <u>texture</u> on <u>each side</u>. It's often used where only <u>one surface</u> is <u>seen</u>, so that only one side needs to be <u>smooth</u> for <u>printing</u>. This board is often used for <u>food packaging</u>.

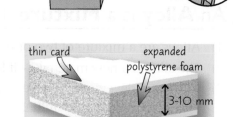

thin card

expanded polystyrene foam

3-10 mm

thin card

<u>Foam core board</u> is made by sandwiching expanded <u>polystyrene foam</u> between 2 thin layers of <u>card</u>. It's <u>stiff</u>, <u>lightweight</u> and the thin outer card layer <u>can be scored</u> (see p.67). It's good for <u>making models</u> and <u>mounting posters</u>.

Foam core board and foil lined board are both laminated (see p.46) and composite materials (see p.39).

<u>Foil-lined board</u> is a board with an <u>aluminium foil lining</u>. It's often used to <u>package food</u> (see p.62) — the foil keeps <u>flavours in</u>, and <u>air</u> and <u>moisture out</u>.

Paper, Board and Timber

Wood is Either **Softwood** or **Hardwood**

Wood comes in two sorts — expensive hardwoods and cheaper softwoods.
Timber refers to sawn chunks of solid wood that are used as building materials.

Softwoods

Softwoods grow in colder climates and are fast growing — this makes
them fairly cheap and readily accessible. The trees have leaves
like needles, are usually evergreen and have cones (e.g. pine).

a knot

1) Pine is yellow with brown streaks. It's quite strong and
   cheap — but knotty, which makes it harder to work with.
   It's used for telegraph poles, fences and cheap furniture.

2) Larch has an attractive yellow to reddish-brown colour.
   It's harder, tougher and more durable than most softwoods.
   It's also resistant to rot, which makes it good for decking,
   cladding the outside of buildings, and fence posts.

3) Spruce is a reddish brown colour. It's hard and has
   a good strength-to-weight ratio — but it's also knotty
   and not very durable. It's used for structural purposes
   both outside and in — aircraft, crates and ship masts.

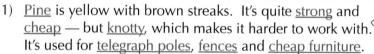

Hardwoods

Hardwoods usually grow in warm climates and are slow growing — so they're generally
more expensive than softwoods. The trees have broad, flat leaves and are usually
deciduous (they lose their leaves in autumn). The wood tends to have a tighter grain and
be denser and harder than softwood, although there are exceptions to this, e.g. balsa.

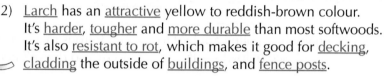

1) Oak is light brown. It's tough, durable and very strong. It also has attractive
   grain markings, especially when quarter sawn (see p.55), and finishes well,
   so it's used in a lot of interior panelling, flooring and furniture.
   However, it does corrode steel screws and fittings.

2) Mahogany is a red-brown colour. It's durable and easy to work with,
   but it's expensive, so it's used for good quality furniture.

3) Beech is pinkish-brown. It's hard enough to resist being dented, and can
   be bent using steam. It's used for chairs and toys.

4) Balsa is a white or tan colour. For a hardwood, it has a very low density and
   is very soft. Its softness makes it easy to cut and shape — this combined with
   a high strength-to-weight ratio, makes it great for modelling (see p.137).

5) Ash has a pale cream colour. It's tough and absorbs shock well, so it's
   used for tool handles and wooden sports equipment (e.g. baseball bats).
   It's also attractive so is used for furniture.

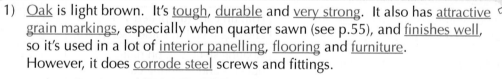

Hardwoods tend to be harder, denser, and pricier than softwoods...

Both types of wood have something that they are well-suited to. Best get learning them.

Metals, Alloys and Polymers

Some metals are pure metals and others (alloys) are mixtures of a metal with another element.
Both types of metal can be classified into two basic groups — ferrous and non-ferrous.

Ferrous Metals Contain Iron

1) These metals or alloys are mostly made up of iron. Because of this, almost all of them are magnetic.

Metal	Properties	Uses
Cast Iron Iron + 2-3.5% Carbon + Silicon	Very strong if compressed, but brittle and not malleable.	Bench vices, car brake disks, manhole covers
Low-Carbon Steel (also known as Mild Steel) Iron + 0.05-0.35% Carbon	Quite strong and cheap, but rusts easily and can't be hardened.	Car bodies, screws, nuts, bolts, nails, washing machines
High-Carbon Steel (also known as Tool Steel) Iron + 0.5-1.5% Carbon	Harder than low-carbon steel, can be hardened. Not as easy to work though and it rusts.	Tools, e.g. chisels, files, saws, drills

Remember, malleable materials are bendy.

2) Protective coatings (e.g. paint, enamel) are often applied to prevent ferrous metals from rusting.
An exception to this is stainless steel (see below) — it's resistant to rust so doesn't need to be coated.

Non-Ferrous Metals Don't Contain Iron

These metals or alloys don't contain iron so don't rust — useful if they're likely to be exposed to moisture.

Metal	Properties	Uses
Aluminium	Lightweight and corrosion-resistant, but expensive and not as strong as steel (although strong for its weight). Hard to join.	Aeroplanes, cans, ladders
Brass 65% Copper + 35% Zinc	Quite strong, corrosion-resistant, malleable, ductile and looks good (nice golden bronze colour).	Electrical parts, door handles, taps
Copper	Relatively soft, malleable, ductile and a very good electrical conductor.	Electrical wiring, pipes.
Tin	Soft, corrosion-resistant, malleable and ductile. Has a low melting point.	Foil, tin cans (tin-plated steel), alloying metal in solder (see p.90).
Zinc	Not very strong, corrosion-resistant.	Coating steel (e.g. on nails, buckets and watering cans).

Alloys can have Really Useful Properties

Alloys are formed when one or more elements are combined with a metal (see p.19). Here are some examples you need to know:

High Speed Steel

1) High speed steel contains iron, more than 0.6% carbon and other metals including chromium, tungsten, and vanadium.

2) It keeps its hardness when heated to high temperatures, so it's used in high speed cutting tools that get hot when used.

Brass (copper + zinc)

1) Brass is harder and stronger than both copper and zinc.

2) It is malleable, ductile and a good electrical conductor (like copper), and is resistant to corrosion (like zinc).

Stainless steel (iron + carbon + chromium + nickel)

1) Cast iron and mild steel (iron + carbon) are strong but rust (corrode) easily.

2) Adding chromium and nickel increases strength, toughness and ductility, and decreases rust.

3) It is used in surgical equipment, sinks and cutlery.

Metals, Alloys and Polymers

Next up, plastics. Plastics are <u>made</u> of <u>polymers</u> (see p.56) but sometimes they just get <u>called</u> 'polymers'. There are <u>two main sorts</u> — <u>thermoforming</u> and <u>thermosetting plastics</u>.

Thermoforming Plastics are Recyclable and Bendy

1) Thermoforming plastics <u>don't resist heat</u> well — they're easily formed into different shapes by <u>heating</u>, <u>melting</u> and <u>remoulding</u>.

2) This means they're <u>easy to recycle</u> — they're <u>ground down</u>, <u>melted</u> and <u>re-used</u>.

Thermoforming Plastic	Properties and Uses
Acrylic (PMMA)	<u>Hard</u>, <u>stiff</u> and <u>shiny</u>. <u>Resists weather</u> well. Can be used to make motorcycle helmet visors, baths, signs, etc. Quite <u>brittle</u> (not tough).
High-density polyethylene (HDPE)	<u>Stiff</u> and <u>strong</u> but <u>lightweight</u>. Used for things like washing-up bowls, baskets, folding chairs and gas and water pipes.
Polyethylene terephthalate (PET)	PET is a <u>polyester</u> that is <u>light</u>, <u>strong</u> and <u>tough</u>. Used to make see-through drink bottles and fibres for clothes.
High impact polystyrene (HIPS)	<u>Rigid</u> and <u>fairly cheap</u>. Used for vacuum forming and fabricating boxes like CD cases or smoke detector casings.
Polyvinyl chloride (PVC)	<u>Quite brittle</u>, <u>cheap</u> and <u>durable</u>. Used for blister packs, window frames, vinyl records and some clothing.
Polypropylene (PP)	Quite <u>tough</u> and <u>flexible</u>. Can be made in a wide variety of <u>bright colours</u>. Used for plastic chairs — its flexibility makes it comfortable.

Thermosetting Plastics are Non-Recyclable and Rigid

1) Thermosetting plastics <u>resist heat</u> and <u>fire</u> (so they're used for electrical fittings and pan handles).

2) <u>Thermosetting plastics</u> (unlike thermoforming plastics) undergo a <u>chemical change</u> when <u>heated</u> and <u>moulded</u> to make a product — they <u>permanently</u> become <u>hard</u> and <u>rigid</u>.

3) This means they're <u>non-recyclable</u> — they <u>can't</u> be melted and reshaped again.

Thermosetting Plastic	Properties and Uses
Epoxy Resin (ER)	<u>Rigid</u>, <u>durable</u>, <u>corrosion-resistant</u> and a <u>good electrical insulator</u>. Used for circuit boards and wind turbine rotor blades.
Urea-Formaldehyde (UF)	<u>Hard</u>, <u>brittle</u> and a <u>good electrical insulator</u>. Used for things like plug sockets and cupboard handles.
Melamine-Formaldehyde (MF)	<u>Strong</u> and <u>scratch-resistant</u>. Used to laminate chipboard and for plates and bowls.
Phenol-Formaldehyde (PF)	<u>Hard</u> and <u>heat-resistant</u>, it's very <u>easily moulded</u> into bottle caps, snooker balls, etc. Also mixed with other materials to form a composite (see p.39).
Polyester Resin (PR)	<u>Hard</u>, <u>stiff</u>, <u>cheap</u> and a <u>good electrical insulator</u>. Added to glass fibres to form glass-reinforced plastic (see p.39), e.g. for kayaks. Waterproof, so used in shower stalls and garden furniture.

Make sure you know the difference between thermoforming and thermosetting plastics, and can give examples of each.

Thermoforming plastics — heat it up and form something new...

It's quite easy to get thermoforming and thermosetting plastics mixed up. Thankfully, I've thought of a nice way to remember it. Thermo**form**ing plastics can be melted and remoulded to **form** new products. Thermo**set**ting plastics can't be remoulded — once set, they're **set** permanently.

Textiles

Textiles, also called fabrics, can be made from natural or synthetic fibres.

Natural Fibres come from Natural Sources

1) Natural fibres can be harvested from plants and animals — cotton comes from plants and wool from sheep.
2) These are renewable resources (i.e. you can always produce more of them). They are also biodegradable (can be broken down by microorganisms in the ground when disposed of) and recyclable — natural fibres are therefore fairly sustainable.
3) In general, natural fibres are absorbent and strong when dry, but have poor resistance to biological damage, e.g. from moths and mould.

Fibre	Properties			Used in these fabrics	Uses
	Appearance	Good points	Bad Points		
Cotton	Smooth	Strong, hard-wearing, absorbent, comfortable to wear, feels quite cool in hot weather, easy to wash & add colour to, doesn't cause allergies, non-static, fairly cheap.	Creases easily, high flammability, poor elasticity, can shrink when washed, dries slowly.	E.g. denim, corduroy, calico	Jeans, T-shirts, blouses, soft furnishings.
Wool	Soft or coarse	Warm, absorbent, good elasticity, low flammability, crease-resistant, available in lots of fabric weights.	Can shrink when washed, dries slowly, can feel itchy, fairly expensive.	E.g. knitted fabrics, Harris Tweed, gabardine, jersey, felt	Suits, jumpers, coats, dresses, carpets.
Silk	Very smooth and glossy.	Smooth, resistant to shrinking/stretching, absorbent, good drape, low flammability, comfortable to wear, lightweight, sun-resistant.	Creases easily, might not wash well, weak when wet, expensive.	E.g. organza, chiffon, satin	Lingerie, underwear, dresses, shirts, ties.

Synthetic Fibres are Man-Made

1) Synthetic fibres are made from polymers (long chain molecules). These molecules come mainly from coal and oil — non-renewable fossil fuels. Synthetic fibres are therefore much less sustainable than natural fibres.
2) Synthetic fibres can be given many different properties. In general, they're resistant to biological damage, and can be changed by heating to form different shapes and textures. However, they're not very absorbent, so they can be hard to dye.

Fibre	Properties			Used in these fabrics	Uses
	Appearance	Good points	Bad Points		
Elastane	Soft	Extremely elastic (stretches up to 7 times its length), strong, hard-wearing, lightweight, keeps its shape well, resists sun/biological damage.	Not absorbent, high flammability, not biodegradable.	E.g. LYCRA®	Sportswear, underwear, combined with other fibres to add stretch.
Polyester	Smooth, can have many different finishes.	Strong (even when wet), hard-wearing, low flammability, good elasticity, cheap, resists creasing, dries quickly, resists biological damage.	Not absorbent, not biodegradable, damaged by strong acids, melts as it burns (so harmful if clothes catch fire).	E.g. DACRON®	Sportswear, bed sheets, curtains, cushions, padding, tablecloths.
Polyamide	Can have many different finishes.	Strong, hard-wearing, warm, good elasticity, crease resistant, resists biological damage, fairly cheap.	Not very absorbent, damaged by sunlight, melts as it burns.	E.g. nylon	Sportswear, socks, tights, furnishings, carpets.

Textiles

Fabrics are Made with **Fibres** that are **Spun** into **Yarns**

Before we talk about fabrics, you need to know a bit more about how they are made...

1) Yarns are threads that are woven or knitted to make fabrics.

2) Yarns are made of fibres (tiny hairs). These come in either short lengths (staple fibres) or longer lengths (filaments).

3) Filaments can be spun (twisted together) or used as they are. Staple fibres are spun to produce yarns.

Yarns made from filaments are smooth...

...while yarns made from staple fibres tend to be hairier.

4) Yarns are available in different thicknesses. A 1-ply yarn is a single yarn, a 2-ply yarn is two yarns twisted together, a 3-ply yarn is three yarns twisted together... and so on...

Fibres and **Yarns** are Used to Make **Fabrics**

There are three main ways of turning fibres or yarns into a fabric — they are weaving, knitting or bonding. There's more about each of these types of fabric on the next page.

1 WOVEN FABRICS are made by interlacing (crossing alternately over and under) two sets of yarns.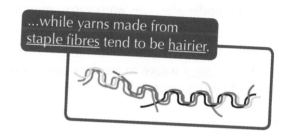

2 KNITTED FABRICS are made by interlocking one or more yarns together using loops. The loops trap air, so they insulate. They stretch more than woven fabrics.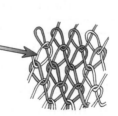

3 NON-WOVEN FABRICS are layers of fibres (not yarns) held together by bonding or felting. They don't fray, and can be cut in any direction — which means there's little waste when laying out patterns (see p.150). However, they don't stretch and aren't very strong.

Natural or synthetic — the choice is yours...

There are absolutely loads of fibres and their properties listed here. To help you to memorise them, you may find it useful to picture an item of clothing that's made from each fibre. For example, we all know that jeans are pretty strong and hard-wearing, but they're also easy to wash and comfortable — these properties are due to the cotton fibres they're made from.

Textiles and Manufactured Boards

It's time to go into a bit more <u>detail</u> about how each type of fabric is <u>made</u>. Here we go...

Fabrics can be made from **Woven Yarns**...

1) Different <u>woven fabrics</u> are made by interlacing <u>two</u> sets of yarns — the <u>weft</u>, which travels from <u>right to left</u> (<u>purple</u> in the diagram below) and the <u>warp</u>, which travels <u>up and down</u> the weave (<u>green</u> in the diagram below).

- The <u>simplest</u> weave is the <u>plain weave</u> — the weft yarn passes over and under <u>alternate</u> warp yarns, making it <u>unpatterned</u>.
- It's <u>hard-wearing</u> — <u>strong</u> and holds its <u>shape</u> well. It has a <u>smooth</u> finish (making it good for printing on).
- It's the <u>cheapest</u> weave to produce and is used to make loads of fabrics, especially <u>cotton-based</u> ones.

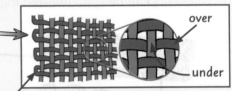

over

under

The <u>edge</u> of a woven piece of fabric, where the weft yarns wrap around the warp yarns, is called the <u>selvedge</u>.

2) Fabrics are woven using <u>looms</u>. Smaller looms are <u>hand-operated</u> and used to create small pieces. However, <u>industrial looms</u> are used to <u>produce woven yarns on a large scale</u>. They are <u>computer operated</u> and much faster than weaving by hand, meaning that fabrics can be <u>mass produced</u> at <u>high speed</u>.

3) Woven fabrics are used for <u>shirts</u>, <u>upholstery</u> and <u>trousers</u> etc.

... **Knitted Yarns**...

There are a <u>two</u> main types of <u>knitted fabrics</u> — <u>weft-knitted</u> and <u>warp-knitted</u>:

Weft-knitted

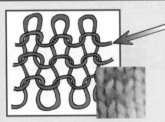

1) The yarn runs <u>across</u> the fabric, making interlocking <u>loops</u> with the <u>row</u> of yarn <u>beneath</u>.
2) These fabrics <u>stretch</u> and can <u>lose their shape easily</u>.
3) If the yarn breaks it can <u>unravel</u> and make a 'ladder'.
4) Weft-knit fabrics can be produced by <u>hand or machine</u>.
5) They're used for <u>jumpers</u>, <u>socks</u> and <u>T-shirts</u>.

Warp-knitted

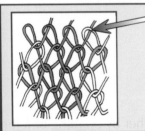

1) The yarns run "<u>up</u>" the fabric, in <u>loops</u>, which <u>interlock vertically</u>.
2) They're <u>stretchy</u> but still <u>keep their shape</u>.
3) These fabrics are <u>hard to unravel</u> and are less likely to ladder.
4) They're made by <u>machine</u> — and the machines can be expensive.
5) <u>Tights</u>, <u>swimwear</u>, <u>fleeces</u>, and some <u>bed sheets</u> are all made from warp-knitted fabrics.

... or **Non-Woven Fibres**

Bonded Fabrics

1) These are "<u>webs</u>" of synthetic fibres that are <u>glued</u>, <u>needle-punched</u>, <u>stitched</u>, or <u>melted</u> together.

Stitch-bonded non-woven fabric

fibres

stitching

2) They're used for <u>interfacing</u> (putting in between layers of fabric for thickness, e.g. in collars — see p.46), <u>artificial leathers</u> and <u>disposable cloths</u> (e.g. medical masks).

Felted Fabrics

1) <u>Felting</u> is an older way of making non-woven fabric. Felt is made by combining <u>pressure</u>, <u>moisture</u> and <u>heat</u> to <u>interlock</u> a <u>mat</u> of <u>wool fibres</u>.
2) Felt can be used for <u>carpet underlay</u>, <u>craft material</u>, <u>hats</u>, <u>jewellery</u> and <u>snooker table coverings</u>.

Textiles and Manufactured Boards

Fibres can be Combined by **Blending** or **Mixing**

Fabrics made from a <u>combination</u> of different <u>fibres</u> can give you <u>better</u> properties.

There are <u>two ways</u> of combining fibres to get fabrics with combined properties — <u>blending</u> fibres within yarns, and <u>mixing</u> different yarns (where each yarn is made from one type of fibre). Here's how:

1) A <u>BLEND</u> is when two or more <u>different fibres</u> are combined to produce a <u>yarn</u>.

2) This mixed yarn is then <u>woven</u> or <u>knitted</u> to make a <u>blended fabric</u>.

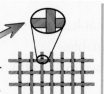

1) A <u>MIX</u> is when a fabric is made up of two or more <u>different types of yarn</u>.

2) The <u>two</u> different types of yarn can then be <u>knitted</u> or <u>woven</u> together to make a <u>mixed fabric</u>.

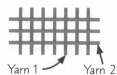

Yarn 1   Yarn 2

Combining <u>cotton</u> and <u>polyester</u> fibres is one of the most common <u>blended</u> fabrics. The resulting fabric:

- is even <u>stronger</u> and remains <u>hard-wearing</u>
- is <u>less absorbent</u> — so dries more quickly
- is <u>soft</u> and <u>comfortable</u>

- resists <u>creasing</u> (is easier to iron)
- <u>doesn't shrink</u> easily
- BUT is <u>highly flammable</u>

Boards can be **Manufactured**

<u>Processed</u> pieces of <u>wood</u> can be combined with <u>glue</u> and <u>compressed</u> into <u>panels</u> — this forms new materials called "<u>manufactured boards</u>" or "<u>manufactured timbers</u>" (see p.55). Here are a few examples you should know about:

Medium Density Fibreboard (MDF)

1) MDF is made from <u>tiny fibres</u> of <u>softwood timber</u> held together by <u>glue</u>.

2) As it's just bits of timber stuck together, it has <u>no natural grain</u>.

3) It's pretty <u>cheap</u>, <u>dense</u> and has a <u>smooth uniform surface</u> that <u>takes paint</u> and <u>other finishes</u> well. However, it's <u>porous</u> so can be <u>damaged by moisture</u>.

4) It's often used for <u>shelves</u> and <u>flat-pack furniture</u>.

Plywood

1) Plywood is made up of several <u>layers</u> of softwood or hardwood, glued together with their <u>grain</u> at <u>right angles</u> to one another (see p.46).

2) That structure makes it <u>very strong</u> for its weight and thickness, compared with solid wood.

3) It's a very popular manufactured board, used for <u>building</u> and <u>furniture</u>.

Chipboard

1) Chipboard is made by compressing <u>wood chips</u>, <u>shavings</u> and <u>sawdust</u> together with glue. It's usually used with a <u>veneered surface</u>.

2) It's <u>cheap</u> but <u>not very strong</u>. It's also <u>absorbent</u>, so can be easily <u>damaged by moisture</u>.

3) Chipboard is often used in <u>cheap self-assembly furniture</u>.

Mixing and blending sounds more like food technology to me...

Mixing and blending fibres is actually really useful — you get a new set of properties to make use of.

Warm-Up and Worked Exam Questions

Materials have properties that make them suited to a specific function or use. Let's hope your mind has a sponge-like absorbency for taking on the information in this section. Try these questions to find out...

Warm-Up Questions

1) Plastic is used for the casing of plugs. Suggest one property that makes plastic suitable for this.
2) What happens when ink bleeds?
3) State two properties of copper that make it suitable for use in electrical wiring.
4) State two reasons why natural fibres are sustainable.
5) Give an advantage that warp-knitted fabrics have over weft-knitted fabrics.

Worked Exam Questions

1   **Figure 1** shows a flat-pack cupboard that has been painted white.

Name a manufactured board that would be suitable for making the flat-pack cupboard shown in **Figure 1**. Give **two** reasons for your answer.

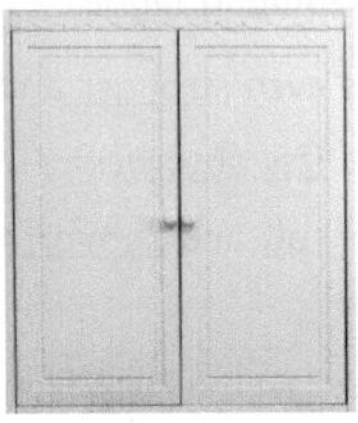

Name: MDF

Figure 1

Reasoning: MDF is a cheap material, so it's suitable for flat-pack furniture, which

is generally sold cheaply. It has a smooth uniform surface, so it can take paint well.

[3 marks]

2   High speed steel is an alloy that can be used to make high speed drill bits.

Explain why high speed steel is a suitable material for this purpose.

High speed steel keeps its hardness when heated to high temperatures. This makes it a

suitable choice for a drill bit because drilling materials at high speed generates heat.

It is important that the drill bit remains hard, because drills can only
operate on materials that are softer (less hard) than the drill bit.   [2 marks]

3   Fabric for a school cardigan is to be made from 100% wool.

a)   State **two** properties of wool that make it suitable for a cardigan.

1.   It is warm.

2.   It is crease-resistant.   You could have given other properties such as having a
good elasticity or being available in lots of fabric weights.   [2 marks]

b)   Give **two** reasons why wool might not be the best fibre to use for a school cardigan.

1.   It can shrink when washed.

2.   It can feel itchy to the wearer.   Other reasons include that it is fairly expensive and it dries slowly.

[2 marks]

Exam Questions

1   Which **one** of the following is a type of natural fibre?

    A   Elastane   ☐

    B   Cotton   ☐

    C   Polyamide   ☐

    D   Polyester   ☐

[1 mark]

2   State **one** property and **one** use for balsa wood.

Property: ....

Use: ....

[2 marks]

3   Fibres can be combined by blending or mixing to give fabrics that incorporate the properties of the different fibres used. Explain the difference between a blended fabric and a mixed fabric.

....

....

[2 marks]

4   Suggest a plastic that would be suitable for making each of the products shown in **Figures 2** and **3**. Give **one** reason for each answer.

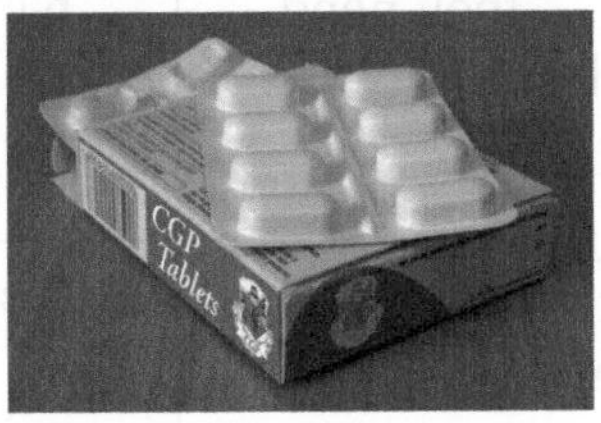

Figure 2

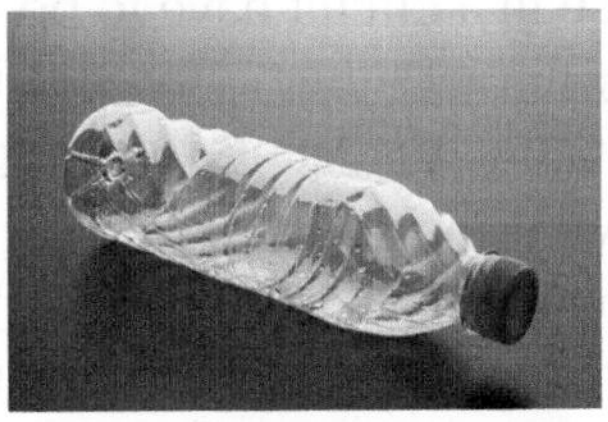

Figure 3

  a) **Figure 2**

Plastic: ....

Reason: ....

....

[2 marks]

  b) **Figure 3**

Plastic: ....

Reason: ....

....

[2 marks]

Electronic Systems

A system is a collection of parts that work together to do a particular function. Electronic systems are made up of components that are connected to form a circuit — all electronic products have one...

All Systems have an Input, a Process and an Output

1) A system can be broken down into three 'blocks' — input, process and output.

2) A signal (e.g. electricity, movement, light) passes from one block to the next.

3) Each block changes the signal in some way.

It might be easier to understand these blocks using an example from a mechanical system (see p.34). A bicycle is a good example:

Input

Process

Output

1) The turning of the cyclist's legs is the input signal to the pedals (the input block).

2) This is then processed in the chain and sprockets (the process block).

3) The output block is the rear wheel, which propels the bike forward.

Sprockets are toothed wheels that are connected by a chain.

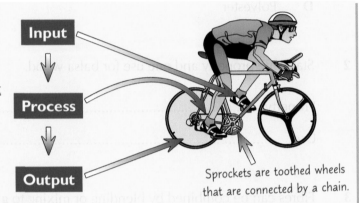

Electronic Systems Involve Circuits

1) Circuits are made up of components that are joined together by wires.

2) Lots of electronic systems, however, make use of printed circuit boards (PCBs). These are boards with thin copper tracks (instead of wires) connecting components in the circuit — they're used to reduce the size and manufacturing cost of electronic systems.

3) The materials you use in a circuit have to be conductors — they need to let electricity flow through them. Copper is used for the wire or tracks because it's ductile, malleable and a good conductor.

4) Circuits can be drawn using a circuit diagram — symbols are used to represent components, and the straight lines linking them represent how they're connected up (e.g. with wires).

5) Voltage from a power cell (e.g. a battery) or the mains pushes the electric current around a circuit.

Copper tracks

Component	Symbol
Battery	⊣⊢⊣⊢
Resistor	▭
Switch	open closed
Buzzer	⊔
Bulb	⊗
Speaker	◁

6) Resistors can be used to reduce the current in a circuit so you don't damage delicate components (such as a light bulb). Resistance is measured in ohms (Ω). A larger resistance means less current in the circuit.

Colour-coded stripes show the resistance.

7) In electronic systems, the input, process and output blocks are the circuit components, e.g. a switch, buzzer, lamp etc.

For example — an electronic egg timer:

Input signal	Input block	Signal	Process block	Signal	Output block	Output signal

SWITCH ⟹ **TIMER** ⟹ **BUZZER**

You press the switch...	...which completes the circuit and starts the timer...	...which adds a set time delay to the system (see p.33). When this is up, it passes on the signal to the buzzer...	...which makes loads of noise.

Electronic Systems

Input Devices Change the Electrical Current in a Circuit

1) Input devices <u>receive</u> an <u>external</u> signal — this is what <u>triggers</u> a <u>system</u> to work.

2) They're often <u>sensors</u> — <u>components</u> that <u>convert</u> a specific type of <u>energy</u> into an <u>electrical signal</u>.

3) There are various types of <u>input device</u> that you can use in an electronic system. These include <u>switches</u> and <u>variable resistors</u>.

Switches Turn a Circuit On and Off

1) Turning a switch on <u>completes</u> a circuit. This <u>allows current to flow</u>.

2) There are many different types of switches — common ones include <u>toggle</u>, <u>push</u> and <u>slide</u> switches.

Toggle switch —
flick to one position to
complete the circuit.

Push switch —
push down and hold to
complete/break the circuit.

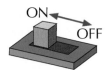

Slide switch —
slide to one side to
complete the circuit.

Variable Resistors Can Detect Changes in External Conditions

1) Variable resistors <u>change the resistance</u> of the circuit, often based on an <u>external factor</u> such as temperature.

2) This alters the <u>current</u> in the wires, which gets passed on to the <u>process block</u>.

3) Here are some <u>common types</u> of variable resistor:

Thermistors

1) <u>Thermistors</u> detect changes in <u>temperature</u> — in <u>hot</u> conditions, the <u>resistance falls</u> and <u>lots</u> of <u>current</u> can flow through them.

2) They're used in <u>central heating systems</u> that <u>turn off</u> when the room reaches a <u>certain temperature</u>.

Light-dependent resistors

1) <u>Light-dependent resistors</u> (LDRs) are <u>light sensors</u> — they detect changes in <u>light levels</u>.

2) In <u>brighter light</u>, the <u>resistance</u> of an LDR <u>falls</u> — this allows circuits to be turned <u>on or off</u> when it gets <u>light or dark</u>.

3) For example, LDRs are often used in <u>automatic night lights</u> — the light is turned <u>on</u> when it gets <u>dark</u>.

Pressure Sensors

1) <u>Pressure sensors</u> detect changes in <u>pressure</u> in a <u>system</u>.

2) Depending on the <u>type</u> of pressure sensor, resistance can either <u>increase or decrease</u> as pressure increases.

3) They're often used in equipment where <u>gases or liquids</u> are <u>monitored</u>. In <u>cars</u>, pressure sensors are used to detect if there's a <u>leak</u> in the <u>fuel system</u> — a leak would cause the pressure to <u>fall</u>, and this would <u>trigger</u> a <u>warning light</u> to come on.

In a sealed container
or system, the more
gas or liquid there is,
the higher the pressure.

Switches and variable resistors can trigger a system into action...

Input devices are exactly what they say on the tin — they allow an input into a system. The system then has to make a decision based on the input (or inputs), but we'll get on to this on the next page...

Electronic Systems

Now it's time for the juicy part of electronic systems — processing the inputs and producing outputs.

Process Devices Receive Inputs and Decide the Outputs

Process devices make all the decisions in an electronic system — they process the input information and use it to determine the output. Devices called integrated circuits can carry out processing in electronic systems.

Integrated Circuits Combine Many Electronic Components

1) Integrated circuits (ICs) or 'chips' are tiny, self-contained circuits. They can have billions of components built in, so they can get very complicated.

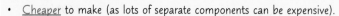

2) ICs help to simplify electronic systems by reducing the number of separate components that are needed in a circuit. This means that systems with ICs are:

   • Cheaper to make (as lots of separate components can be expensive).
   • Smaller — this is particularly important in portable electronic products, e.g. mobile phones.
   • Use much less power (because of their size).

3) You can treat ICs as a single process block.

There are loads of different types of integrated circuit — microcontrollers are just one type of IC.

Microcontrollers are a Special Type of Integrated Circuit

1) A microcontroller is a mini computer on a chip — it has a processor, memory, and one or more inputs and outputs.

2) They're found in pretty much all electronic products, e.g. microwave ovens, washing machines etc.

Advantages
1) Microcontrollers can do the jobs of multiple ICs — this simplifies the system even further.
2) Unlike other ICs, most can be reprogrammed for a different use.

Disadvantage
Microcontrollers are more expensive than other ICs.

A commonly used microcontroller is a PIC, a programmable intelligent computer (see p.115).

Microcontrollers are Programmed for a Certain Task

1) Microcontrollers need to be programmed (given instructions) to perform a particular function — without this, they won't do anything at all. The program is stored in the microcontroller's onboard memory.

2) Often the programming of microcontrollers is carried out when they are manufactured and the program isn't altered after this. However, it is possible to program microcontrollers yourself.

3) There are several microcontroller boards (a PCB with a microcontroller along with other components) that can be programmed for use in a variety of projects — e.g. Arduino boards and the BBC micro:bit.

4) Programming can involve either:

Different shaped boxes represent different stages in a flowchart.

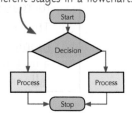

   • Writing a series of commands using a programming language — examples include BASIC and Scratch.
   • Creating a flowchart telling the microcontroller exactly when to start, how to make decisions based on input signals, when to add a time delay to the process, what to output, and when to stop operating.

5) This is done using specialist computer software such as PICAXE (which uses BASIC) and Logicator (which uses flowcharts) or online tools such as the BBC micro:bit website. The instructions are transferred to a microcontroller so they can be used.

6) Microcontrollers can be programmed for a range of different tasks. You need to know about them being programmed as timers, counters and for decision-making — see next page.

Electronic Systems

Microcontrollers as **Timers** and **Counters**

1) Timers and counters have similar functions — they can both be involved in timing in circuits. They both work by responding to pulses (peaks of higher voltage) in circuits.

2) Timers are often used to add a time delay to a process and make sure parts of a program are running when they're supposed to — they can generate a pulse after a particular length of time has passed.

3) Counters differ from timers because they count pulses of voltage produced by an input device.

4) Microcontrollers usually contain timers and counters inside them, which allow the programs they run to include timing and/or counting functions.

Microcontrollers as Timers

- Microcontrollers controlling flashing lights, like car indicators, use timers to set the time that the light is on and off for.
- Microcontrollers in microwaves use a timer to set the length of time that the oven cooks for.

Microcontrollers as Counters

Most modern cars have a dial on the dashboard that shows how fast the engine is turning — the number of revolutions per minute. A microcontroller is used to count the number of revolutions.

Microcontrollers as **Decision-Makers**

1) Logic gates make decisions in ICs based on a collection of inputs.

2) Inputs must be digital — they're either on (1) or off (0).

3) The names of logic gates are, well, pretty logical...

- OR gates have 2 inputs: if one input or the other is on (or both), then the output is on. An automatic door is a good example — the door will open if a sensor detects someone on the outside or inside.
- AND gates have 2 inputs: both inputs need to be on for the output to be on. Tumble dryers won't start until both the start button is pressed and an input sensor has detected that the door is closed.
- NOT gates have 1 input: if the input is on, the output is not on. It reverses the input. Pressing an emergency stop switch will stop the output — on an escalator, this would cause it to stop moving.

Output Devices Include **Lamps**, **Buzzers** and **Speakers**

1) The output device you use in a system depends on what your product has to do. It determines how the system responds to the input signal.

2) There are absolutely loads of output devices... here are a few you need to know for the exam:

LEDs (light-emitting diodes) and lamps (e.g. bulbs) turn electricity into light, e.g. security lights.

Buzzers make a noise. They're used in alarm clocks to wake you up in the morning.

Speakers turn electrical signals into sound, e.g. in headphones.

System triggered, decisions made, output activated, job done...

Don't be put off if there's an unfamiliar system in the exam. It's just a case of applying what you already know about Tinputs, processes, outputs, etc. Make sure you read the question carefully, and if you're really stuck, move on and come back to it if you have time at the end.

Mechanical Systems

Electrical systems aren't the only type of system you need to know about. There are also mechanical systems.

Mechanisms Change the Magnitude and Direction of Force

1) All mechanical systems have mechanisms which transform an input motion and force into an output one.

2) They're designed so you can gain a mechanical advantage from using them — they make something easier to do. This often involves changing the magnitude (size) and direction of the applied force.

3) Some mechanisms change one type of motion into another.

4) There are various types of motion:

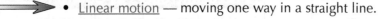
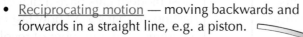

- Linear motion — moving one way in a straight line.
- Reciprocating motion — moving backwards and forwards in a straight line, e.g. a piston.
- Oscillating motion — moving backwards and forwards in an arc, e.g. a swing.
- Rotary motion — moving in a circle, e.g. a wheel.

For example, a car jack lets you lift up a car — a job you couldn't do without it. You turn a handle (rotary motion) with a relatively small force and the car is lifted upwards (linear motion) by a large force.

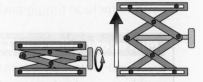
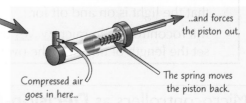

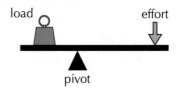

...and forces the piston out.

Compressed air goes in here...

The spring moves the piston back.

Levers Make It Easier to Move and Lift Things

Levers move and lift loads by rotating about stationary points called fixed pivots. There are three main types of lever that you need to know about.

First Order Levers Have a Pivot in the Middle

1) First order levers have the pivot between the effort and the load. Pushing down on your side of the lever lifts the load up at the other end — the lever has rotated about the pivot.

2) If the load is closer to the pivot than the effort, a large load can be lifted using a smaller effort — the lever gives a mechanical advantage.

3) As you move the pivot closer to the load it becomes easier to lift — the magnitude of the effort required decreases.

load   effort

pivot

First order levers can make it much easier to lift heavy loads.

Second Order Levers Have the Load in the Middle

1) Here the pivot is at one end of the lever and the effort is at the other end. The classic example is a wheelbarrow.

2) The closer together the pivot and load are, the easier it is to lift.

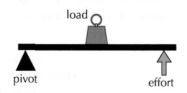

load

pivot   effort

effort

pivot

load

Third Order Levers Have the Effort in the Middle

1) In a third order lever the effort is in between the load and the pivot.

2) Third order levers can be things like fishing rods, cricket bats and garden spades.

3) Moving the effort and pivot further apart makes it easier to move or lift the load.

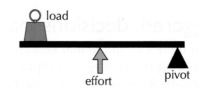

load

effort   pivot

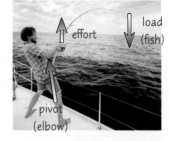

load (fish)

effort

pivot (elbow)

Mechanical Systems

Linkages Connect Different Parts of a Mechanism

Levers can be connected together to form linkages. Simple linkages can change the magnitude of the force and the direction of motion. Here are two examples:

Push/Pull Linkages

1) Push/pull linkages use two fixed pivots.

2) The input and output motions of the linkage are in the same direction. The motion of the link arm is in the opposite direction.

3) Here, each fixed pivot is in the centre of an arm. Changing the position of these pivots will change the magnitude of the output force — often the easiest way to see this is by making a model.

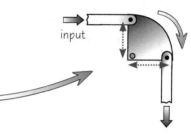

input   output

moving pivot

fixed pivots

link arm

A moving pivot can move in space whilst allowing levers to rotate. (Remember, fixed pivots remain stationary in space.)

Bell Crank

1) A bell crank changes the direction of a force through 90°.

2) The magnitude of the output force can be changed by moving the fixed pivot so it's not an equal distance between the two moving pivots.

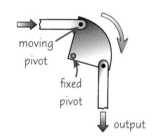

input

moving pivot

fixed pivot

output

Gear Trains Transmit or Change Rotary Motion

1) Gears are toothed wheels that interlock. They transfer motion from one part of a machine to another.

2) A gear train is where two or more gears are linked together. They can be used to change the direction of motion or change the magnitude of the input force (this is to do with the increase or decrease in the speed of rotation that can be generated by linking different-sized gears). Here are some examples:

The driver gear, turned by hand or a motor, turns the driven gear. Both will turn in opposite directions.

Driver — Driven

If you use a third gear (called an idler), the driver and the driven gears will both turn in the same direction. The size of the idler won't alter the speed of the other two gears.

Idler

If linked gears are different sizes, the smaller gear (i.e. the one with fewer teeth) will turn faster. The relationship between the driver and the driven gears can be described using a gear ratio. The size of this ratio describes how much the mechanism changes the speed of the gears from the input speed to the output speed. Here's an example of how gear ratios can be used in calculations:

$$\text{Gear ratio} = \frac{\text{no. of teeth on the driven gear}}{\text{no. of teeth on the driver gear}}$$

$$\text{Output speed} = \frac{\text{speed of driver gear (input speed)}}{\text{gear ratio}}$$

A gear train is made up of a driver gear with 10 teeth and a driven gear with 20 teeth. The driver gear is rotating at 500 rpm (revolutions per minute). Calculate the output speed of the gear train.

Gear ratio = 20 ÷ 10 = 2/1 or 2:1 or 2
Output speed = 500 ÷ 2 = 250 rpm

Driver 10 teeth   Driven 20 teeth

Mechanical Systems

Rotary systems are mechanical systems that work by using rotary motion. Gear trains on the previous page are an example of a rotary system but there are a few more that you need to know about...

Pulleys Help to Lift a Load

1) A simple pulley is made up of a wheel with a grooved outer edge and a cable, rope or belt (see below) that sits in this groove.

2) Pulleys make lifting a load easier.

3) One pulley on its own changes the direction of the force required. The same amount of force is needed but pulling down might be easier than lifting something up.

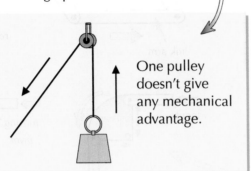

One pulley doesn't give any mechanical advantage.

Cranes use pulley mechanisms.

4) Using two or more pulleys together can change the magnitude of the force too — they can make things feel a lot lighter than they actually are (if you set them up correctly). For example, one fixed pulley and one moving pulley (a block and tackle) will mean you only need half the force to lift a load.

Simple block and tackle

Weight can be lifted using half the effort.

Belt Drives Transfer Movement

1) A belt drive transfers movement from one rotating shaft to another.

2) Belt drives are used in pillar drills. They're also used in products such as washing machines (see p.112).

3) In a pillar drill, a flexible belt joins two separate pulley wheels — this links the motor to the drill shaft.

4) The belt can be put in different positions to make the drill turn faster or slower. This works in a similar way to gears (see previous page) — if the wheels are of different sizes, the smaller wheel will spin faster.

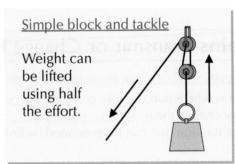

Belt

motor (driver)

drill shaft (driven)

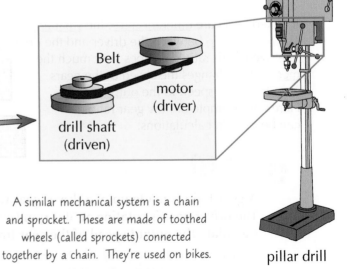

A similar mechanical system is a chain and sprocket. These are made of toothed wheels (called sprockets) connected together by a chain. They're used on bikes.

pillar drill

Mechanical Systems

5) The <u>diameter</u> (width) of the wheels can be used to <u>calculate</u> the <u>velocity ratio</u> — how <u>fast</u> the <u>driven wheel</u> will spin relative to the <u>driver wheel</u>:

$$\text{Velocity ratio} = \frac{\text{diameter of the driven pulley wheel}}{\text{diameter of the driver pulley wheel}}$$

6) You can use another formula to calculate the <u>output speed</u> of the <u>driven</u> pulley wheel.

$$\text{Output speed} = \frac{\text{speed of driver pulley wheel (input speed)}}{\text{velocity ratio}}$$

This formula is very similar to the one we saw for gears on page 35.

EXAMPLE:

A belt drive is made up of a driver wheel with an 30 mm diameter and a driven wheel with a 90 mm diameter. The driver wheel is rotating at 600 rpm. Calculate the output speed of the driven wheel.

Velocity ratio = 90 ÷ 30 = 3/1 or 3:1 or 3   Output speed = 600 ÷ 3 = 200 rpm

Rpm means revolutions per minute — it's a measure of the speed of rotation.

Cams Change **Rotary Motion** to **Reciprocating Motion**

1) A <u>cam mechanism</u> has <u>two main parts</u> — the <u>cam</u> and the <u>follower</u>.

2) Cams come in many different shapes and sizes. They always <u>rotate</u>.

3) The <u>follower</u> rests on the cam as it rotates, and <u>follows its shape</u>. It may have a small wheel to reduce <u>friction</u>. The follower moves <u>up and down</u> (reciprocating motion) as the cam turns.

WHEEL FOLLOWER

CAM

rotates

Here are a few basic <u>cam shapes</u>:

1 <u>CIRCULAR CAM</u> (also called <u>offset</u> or <u>eccentric</u>) — produces a <u>uniform reciprocating motion</u>.

The circular cam rotates about an off-centre pivot...

... which causes the follower to move up and down.

2 <u>SNAIL CAM</u> — For half a turn the follower will not move, then it will <u>gently rise, and then suddenly drop</u>. It will only work in <u>one direction</u>.

Rotates in one direction only

3 <u>PEAR CAM</u> — Again for half a turn the follower will not move, then it will <u>gently rise and fall</u>.

4 <u>FOUR-LOBED CAM</u> — Has four lobes (bits that stick out). For <u>each turn</u> of the cam the follower will <u>rise and suddenly fall four times</u>. This cam shape will also only work in <u>one direction</u>.

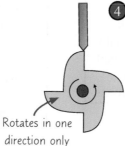

Rotates in one direction only

4) Changing the <u>size</u> or <u>shape</u> of the <u>cam</u> can be used to <u>change</u> the <u>magnitude</u> of the <u>output</u> motion (the <u>reciprocating</u> motion of the <u>follower</u>).

REVISION TIP

If only there was a mechanism to make revision easier...

We've covered loads of mechanisms over the last few pages. But fear not — take them one at a time and make sure you understand what's going on in the diagrams.

Developments in New Materials

New materials are being developed all the time. They can have some seriously snazzy properties too...

Modern Materials are... well, Modern

1) Modern materials are normally materials that have been developed with a specific application in mind.
2) They can be developed by inventing new or improved manufacturing processes. For example:

Graphene

1) Graphene is a super-thin layer of graphite — the stuff used in pencil leads.
2) It has some amazingly useful properties — it's incredibly light and strong and is a great conductor of heat and electricity.
3) It was first made in 2004. Since then, several manufacturing processes have been invented and improved to allow it to be made more cheaply.
4) Graphene can be used in modern tennis racquets. Future applications include aerospace, vehicles, water purification and flexible electronics.

Metal Foams

1) Metal foam is a metal (e.g. aluminium) that contains many gas-filled spaces which make the material lightweight. These foams also keep some of the metal's properties too — they're stiff, tough, and strong under compression.
2) Metal foam was developed in the 1940s. Research since has focussed on improving its properties and production methods.
3) These foams are already being used for lightweight car parts and in bone implants.

Titanium

1) Titanium is an extremely corrosion-resistant metal with a high strength-to-weight ratio.
2) A process was developed in the 1930s that allowed titanium to be extracted easily and cheaply enough to be used in the aerospace industry (aircraft and spacecraft) — this is its main use to this day.
3) Titanium is difficult and therefore expensive to machine (e.g. drill, cut, polish). Recent developments in CAD/CAM have helped to reduce machining costs and make it a more usable material, e.g. for bone replacements, dental implants, bikes, ships and armour.

3) Some modern materials can be developed by altering an existing material to perform a particular function:

Liquid Crystal Displays (LCDs)

1) LCDs are used in flat-screen displays that are thin, lightweight and energy efficient.
2) The liquid crystals used in the display are made of a mixture of chemicals. When an electric current is applied, the crystal's shape is modified — this in turn changes the image seen on the screen.
3) LCDs were originally used in calculators and other small displays and were usually just black and grey. They have developed to become high-definition, full-colour displays used as TV and computer screens.

Coated Metals

1) By coating metals with a different material, their properties can be altered.
2) E.g. iron and steel can be galvanised — coated with zinc to prevent rusting (see p.92). Electroplating can also be used to achieve this, e.g. nickel-plated steel is a cheaper, corrosion-resistant alternative to stainless steel in car parts.
3) Anodised aluminium has a coating of aluminium oxide to make the surface harder and more resistant to corrosion (see p.113).
4) Metals (e.g. steel) can be coated with PVC (see p.23) — used for roofing, the PVC makes it corrosion-resistant and can be coloured.

Nanomaterials

1) Nanomaterials are made of tiny particles (nanoparticles). These particles have always existed but our ability to manipulate them for specific purposes is a fairly recent development. Nanomaterial uses include:
2) Carbon nanotubes are tiny carbon cylinders. They have a very high strength-to-weight ratio, and are good conductors of heat and electricity. They can be added to a material to strengthen it without adding much weight (e.g. in tennis racquets). They're also used in electronics.
3) Self-cleaning fabrics have a nanoparticle coating that removes odours and stains upon exposure to light.
4) Antibacterial fabrics use nanoparticles of silver to kill bacteria. They have lots of medical uses, e.g. face masks and dressings. They're also used in anti-bacterial toys and odour-free socks.

Developments in New Materials

Smart Materials React to Their Environment

Smart materials <u>change</u> their <u>properties</u> in response to stimuli, e.g. <u>temperature</u>, <u>light</u>, <u>stress</u>, <u>moisture</u> or <u>pH</u>. They often <u>return</u> to their <u>original state</u> when the stimulus is <u>taken away</u> — the changes are <u>reversible</u>.

1) <u>Shape memory alloys</u> are alloys that '<u>remember</u>' their <u>original shape</u>. They can be easily shaped when cool, but they <u>return</u> to their <u>original shape</u> when <u>heated</u> above a <u>certain temperature</u>.
2) An example of a shape memory alloy is <u>nitinol</u>. If your <u>glasses</u> are made of nitinol and you accidentally <u>bend</u> them (stress), you can just pop them into a bowl of hot water and they'll <u>jump</u> back <u>into shape</u>.

1) <u>Photochromic pigments</u> change colour <u>reversibly</u> in response to <u>light</u>.
2) They can be put into <u>spectacle lenses</u> to make glasses that turn into <u>sunglasses</u> when it's sunny.
3) <u>Photochromic inks</u> can be used to print t-shirts with <u>designs</u> that only <u>show up in sunlight</u>.

1) <u>Thermochromic pigments</u> and inks are used in <u>colour changing</u> products — they react to temperature.
2) When the <u>temperature</u> changes, the product <u>changes colour</u>. The colour <u>changes back</u> when the object returns to its <u>original temperature</u>.
3) They're used in <u>babies' feeding spoons</u> so the parent knows the food isn't too hot, and in <u>novelty mugs</u> and <u>t-shirts</u>.

Blue spoon   gets hot...   ...and becomes pink

Thermochromic ink can be used to print thermochromic plastic sheets, which change colour as the temperature changes. This is used in some thermometers.

Composites are a Combination of Two or More Materials

<u>Composites</u> are made from <u>two or more different materials</u> bonded together. These enhanced materials often have <u>different</u> (more <u>useful</u>) <u>properties</u> than those of the <u>individual materials</u> they're made from. For example:

Composite	Made from	Properties	Uses
Glass reinforced plastic (GRP)	Glass fibres that are coated in a thermosetting plastic resin.	Stronger and tougher than the plastic by itself, heat-resistant, easy to mould into complex shapes.	Boats, kayaks, surfboards, some car bodies, PCBs
Carbonfibre reinforced plastic (CRP)	Carbon fibres that are coated in a thermosetting plastic resin.	Lighter, tougher and stronger than GRP, but more expensive.	Protective helmets, racing cars, sports equipment, laptops, bulletproof vests

Technical Textiles are Made Just to be Functional

<u>Technical textiles</u> are <u>enhanced fabrics</u>. They're <u>designed purely</u> to be <u>functional</u>, rather than <u>look good</u>. E.g:
1) <u>Kevlar®</u> — a <u>synthetic fibre</u> that can be woven into a <u>really strong</u> fabric that's <u>resistant to abrasion</u>. This makes it useful, e.g. in <u>bulletproof</u> vests, clothing for <u>motorcyclists</u>, to reinforce <u>tyres</u>, etc.
2) Some <u>synthetic fibres</u> are designed to be very <u>fire-resistant</u>, e.g. <u>Nomex®</u>. These enhanced fabrics are <u>different</u> to those <u>treated</u> for flame retardance (see p.96), as Nomex® has fire-resistance <u>built into the fibres</u>, so it <u>can't</u> be <u>washed</u> or <u>worn away</u>. They're used in <u>firefighters'</u> and <u>racing drivers'</u> overalls.
3) <u>Micro encapsulation</u> — this is where <u>tiny droplets</u> of a <u>chemical</u> are coated in <u>shells</u> called <u>microcapsules</u>, and embedded in a <u>microfibre fabric</u> (one woven from very thin synthetic fibres). These chemicals have a <u>functional</u> use, e.g. as <u>insect repellent</u>, an <u>odour neutraliser</u>, or a <u>perfume</u>. They're used in products such as <u>antibacterial socks</u> and <u>scented lingerie</u>.
4) <u>Conductive fabrics</u> — use <u>fibres</u> that can <u>conduct electricity</u>. They're used to <u>integrate electronics into clothing</u> and in <u>touchscreen gloves</u>.

New materials are developed to have really useful properties...

EXAM TIP Quite a lot of the exam questions from this section will be multiple choice. These can be tricky, but don't panic — you can always have a go at answering them. If you're struggling, try to rule out any choices that are definitely wrong. Then pick an answer from one of the options left.

Warm-Up and Worked Exam Questions

There's a lot of information squeezed onto these pages, so have another flick through and make sure you know it all like the back of your hand. When you're happy with it all, give these questions a go.

Warm-Up Questions

1) Describe what a printed circuit board (PCB) is.
2) Describe how an AND logic gate works.
3) Describe the motion produced by the cam shown on the right. ———
4) What material are the nanoparticles made of in antibacterial fabrics?

Worked Exam Questions

1   **Figure 1** shows a crowbar being used to remove a nail from a plank of wood. A crowbar is an example of a lever.

Figure 1

a) What type of lever is a crowbar?

   A first order lever.   Remember, to work out what type of lever it is, you need to look at the positions of the load, pivot and effort (see p.34).   [1 mark]

b) Explain why using this type of lever helps to remove the nail.

   It gives a mechanical advantage meaning the nail can be removed

   with a small effort.

   [1 mark]

2   A company is designing a new set of street lights that turn on and off at set times of the day.

a) Explain how a microcontroller could be used to control when the lights turn on and off.

   A microcontroller could be programmed as a timer to add a set time delay to the

   system. For example, if the timer is activated when the light is switched on, it

   could cause the light to stay on for a set time period before switching off.

   [2 marks]

b) The company found that the lights were turning on too early in the summer, and too late in the winter. They decided to change the street light design so that they turn on when the light levels are below a certain value.

   Name a suitable input device for this system. Give a reason for your choice.

   Input device: A light-dependent resistor (LDR).

   Reason: An LDR detects changes in light levels, so it could switch the circuit on

   when it gets dark enough.

   [2 marks]

Exam Questions

1 What type of material is titanium?

   A a modern material ☐

   B a smart material ☐

   C a composite ☐

   D a technical textile ☐

[1 mark]

2 **Figure 2** shows a push/pull linkage. The output force is the same magnitude and in the same direction as the input force.

Which **one** of the following would cause the magnitude of the output force to be different to the magnitude of the input force?

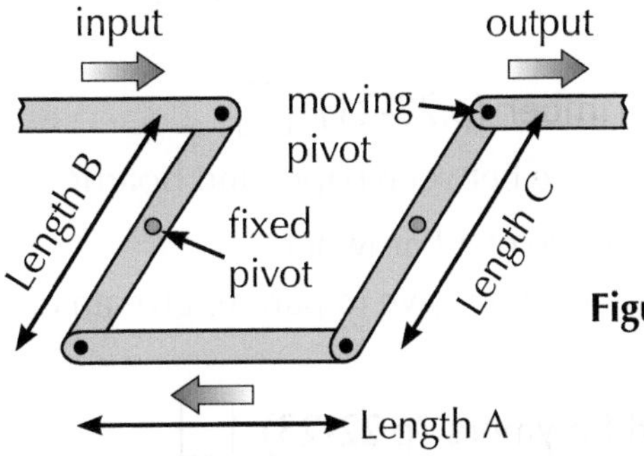

Figure 2

   A Increasing length A. ☐

   B Increasing the magnitude of the input force. ☐

   C Changing the position of the fixed pivots. ☐

   D Increasing length B and length C by the same amount. ☐

[1 mark]

3 A belt drive mechanism transfers rotary motion from a motor to a shaft. Details of the mechanism are given in the table.

Calculate the velocity ratio for this mechanism.

Part of mechanism	Diameter (mm)
Motor (driver)	35
Shaft (driven)	105

....

....

[1 mark]

4 **Figure 3** shows a pair of sunglasses.

   a) State **one** smart material that could be used to make a part of this product.

   ....

   [1 mark]

Figure 3

   b) Explain how the "smart property" of the material you gave in **a)** could be useful to the user of the product.

   ....

   ....

   ....

   [2 marks]

Revision Questions for Section Two

Section Two is done and dusted — try these questions to see how much you can remember.

- Try these questions and tick off each one when you get it right.
- When you've done all the questions for a topic and are completely happy with it, tick off the topic.

Properties of Materials (p.18-19) ☑

1) Describe what is meant by the following properties: a) Fusible   b) Malleable   c) Ductile ☑
2) Absorbent materials soak up moisture. State two other properties of absorbent materials. ☑
3) Metal is used for radiators. Suggest one property that makes metal suitable for this use. ☑

Paper, Board and Timber (p.20-21) ☑

4) What is the difference between paper and board? ☑
5) a) State one example of a hardwood.
   b) For the hardwood you gave in part a), give an example of how it is used. ☑

Metals, Alloys and Polymers (p.22-23) ☐

6) What is a non-ferrous metal? Give an example of one. ☑
7) Explain why stainless steel is useful for making products that are used outdoors. ☑
8) Give two differences between thermoforming and thermosetting plastics. ☑
9) Give a property of melamine formaldehyde that makes it suitable for laminating worktops. ☑

Textiles and Manufactured Boards (p.24-27) ☑

10) Give two examples of synthetic fibres, and state a property of each one. ☑
11) Name the piece of equipment used for weaving. ☑
12) Suggest why chipboard shouldn't be used to make bathroom furnishings. ☑

Electronic Systems (p.30-33) ☐

13) Name the three blocks that make up a system. ☑
14) a) Give three examples of types of variable resistor.
    b) For each resistor named in a), give the external factor that it can detect changes in. ☑
15) Give two advantages of using microcontrollers in a system. ☑
16) Name two output devices and state the type of output signal that they produce. ☑

Mechanical Systems (p.34-37) ☐

17) State an example of a second order lever. ☑
18) A pillar drill operates using a belt drive mechanism. The driver wheel has a diameter of 32 mm and spins at a speed of 1200 rpm. The driven wheel has a diameter of 128 mm.
    a) Calculate the velocity ratio.   b) Calculate the output speed of the system.
    c) Name one other product that uses a belt drive mechanism. ☑

Developments in New Materials (p.38-39) ☐

19) Name two properties of graphene. ☑
20) State two uses of glass-reinforced plastic. ☑

Selecting Materials

When <u>choosing</u> a <u>material</u> to use in a product, there's a lot to consider. There's <u>rarely</u> a <u>perfect</u> material — so you'll have to <u>weigh up</u> which <u>factors</u> are the <u>most important</u> and <u>prioritise</u> those first...

There are Many **Factors** to Think About...

Functionality

1) Probably the most <u>important factor</u> is whether the material is <u>functional</u> — it must have <u>properties</u> that <u>help</u> it to <u>do the job</u> it's <u>supposed to</u> in a product.

2) You need to have an idea of <u>how</u> the product <u>will be used</u> and the <u>demands</u> that will be made on the material. For example, if you were making a road bridge, you'd need materials <u>strong enough</u> to support the weight of vehicles (e.g. steel).

3) You also need to consider how <u>easy</u> the <u>materials</u> are <u>to work with</u>. E.g. you wouldn't want to make a detailed wooden carving out of <u>pine</u> (see p.21) — it's <u>knotty</u>, so would be <u>hard to shape</u>.

Availability

1) You need to consider <u>how easy</u> it is to <u>source</u> (find) and <u>buy</u> the materials you want in a <u>suitable form</u>.

2) <u>Widely available</u> materials are <u>ideal</u> — you can get your hands on them <u>quickly</u> and <u>easily</u>. <u>Less widely available</u> materials are often more <u>expensive</u>. They also may need to be <u>delivered</u> from <u>further away</u>, which adds to the <u>cost</u> and <u>waiting time</u>.

3) A lot of materials are <u>only widely available</u> in <u>stock forms</u> and <u>sizes</u> (see p.77). You should try to use these <u>stock</u> forms and sizes if possible, as it can be very <u>expensive</u> to get them in <u>other forms</u>.

Aesthetics

1) If a material is going to be <u>seen</u>, you want it to <u>look good</u>. You should <u>consider</u>:
   - <u>Colour</u> — e.g. using a thread that's the same colour as a fabric may look better on a garment.
   - <u>Surface finishes</u> — e.g. <u>painting</u>, <u>varnishing</u>, <u>polishing</u>, <u>lacquering</u> etc. all <u>change</u> the <u>appearance</u> of a material, as well as helping to <u>protect</u> it from moisture and dirt.

2) You also need to think about a material's <u>texture</u> — how a material <u>feels</u> to the touch, e.g. rough, soft etc. This is especially important in <u>textiles</u>, when the product will be <u>worn</u>.

3) The aesthetics must <u>appeal</u> to your <u>target market</u>. For example, <u>traditional</u> oak furniture might appeal to <u>older people</u>, whereas younger people may prefer <u>brightly coloured</u> polypropylene.

Cost of Materials

When working out <u>how much</u> you can <u>spend</u> on <u>raw materials</u>, you should always consider:

1) <u>How much</u> you're going to <u>sell</u> the <u>product</u> for. You <u>shouldn't spend loads</u> on expensive <u>materials</u> if you're <u>selling</u> the product <u>cheaply</u> — you'll just <u>lose money</u>.

2) The <u>amount of each material</u> you're going to <u>use</u>. <u>Affordable products</u> can still use fairly <u>expensive materials</u> in <u>small quantities</u>. For instance, a <u>gold-plated</u> photo frame uses far less gold than a <u>solid gold</u> photo frame, so can be <u>made</u> and <u>sold</u> at a much <u>cheaper</u> price.

3) <u>How many products</u> you're making. <u>Mass-produced</u> products are <u>cheaper to make</u> than <u>one-off</u> products, because the <u>manufacturing cost per item</u> gets <u>lower</u> the <u>more you produce</u>. One of the reasons for this is that raw materials can be <u>bought in bulk</u> — because you're buying so much, it allows you to <u>negotiate a discount</u> with the <u>supplier</u>.

Selecting Materials

Environmental Factors

1) Where possible, you should try to use <u>sustainable</u> materials to <u>limit</u> the <u>environmental impact</u> of the product, e.g. materials that are <u>biodegradable</u> or that come from <u>renewable resources</u> (see p.6-8).

2) However, <u>a lot</u> of the <u>materials</u> used in products <u>aren't very sustainable</u>, for example...

3) It's not all bad though — some of these materials can be:
   - <u>Recycled</u> (for a <u>different</u> use), e.g. <u>thermoforming plastics</u> can be <u>melted</u> and <u>remoulded</u> (see p.23).
   - <u>Re-used</u> (for the <u>same</u> use), e.g. plastic <u>bags</u> can be re-used <u>many times</u> before breaking.

4) Using recycled or re-used materials in a <u>product</u> helps to <u>limit</u> the amount of <u>non-biodegradable</u> stuff on <u>rubbish tips</u> and the amount of <u>non-renewable</u> material that needs to be <u>obtained</u>.

Metal	• Comes from non-renewable metal ores • Uses a lot of non-renewable energy to mine it • Mining damages ecosystems
Plastic	• Non-renewable (made from oil) • Not biodegradable
Paper/Wood	• Renewable, but only if trees are replanted to replace the ones cut down • Deforestation (see p.58) damages ecosystems

There's more about the processes involved in sourcing these materials on p.55-58.

Social Factors

1) As a designer, you have a <u>social responsibility</u> — a <u>duty</u> to <u>act</u> in a way that <u>benefits society</u> and the <u>environment</u>. This means you <u>shouldn't select materials</u> based only on which ones will make the <u>most money</u>. Instead, you may need to <u>compromise</u> on factors such as the <u>cost</u> or <u>availability of materials</u>.

2) For example, using <u>Fairtrade</u> materials (see p.132) <u>may cost more</u> than using materials from a less ethical source, but it may be considered worthwhile to have a <u>positive social impact</u> on the <u>lives</u> of the <u>farmers</u> and <u>workers</u> who provide the materials.

3) Using <u>recycled materials</u> in products <u>can cost more</u>, but there are <u>social benefits</u>:
   - To <u>obtain raw materials</u>, non-renewable <u>energy</u> sources are often used, e.g. the <u>burning of fossil fuels</u>. This causes <u>air pollution</u>, which can result in <u>long-term health problems</u> such as <u>lung disease</u> and <u>cancer</u>. Although <u>recycling</u> causes <u>some air pollution</u>, it usually doesn't cause <u>as much</u> as the <u>processes</u> used to obtain raw materials. Recycling can therefore benefit the <u>health of society</u>.
   - Obtaining new materials also requires large amounts of <u>land use</u>, e.g. to <u>mine</u> coal and metal, to <u>drill</u> for onshore oil and gas, and to <u>chop down trees</u>. This can mean a <u>reduction</u> in the <u>quality</u> of <u>life</u> for those who <u>live nearby</u>, <u>relocation of communities</u> and the <u>destruction of livelihoods</u>. Recycling helps to <u>reduce</u> the <u>number of people</u> that are <u>impacted</u> by the <u>materials industry</u>.

Ethical Factors

1) <u>Ethically sourced materials</u> are produced in an <u>environmentally sustainable way</u> that is also <u>fair</u> to <u>workers</u>.

2) Some <u>materials</u> have <u>logos</u> to show they have <u>met</u> certain <u>ethical standards</u>. E.g. the logo of the <u>Forest Stewardship Council</u>® (FSC®) shows that the <u>timber</u> (or paper) in a product has come from <u>responsibly managed forests</u> and/or <u>verified recycled sources</u>.

3) <u>Animal products</u> such as <u>fur</u>, <u>ivory</u> and <u>leather</u> are often viewed as <u>unethical</u>, as animals are <u>killed</u> for the materials. <u>Synthetic materials</u> could be considered as an ethical alternative, e.g. artificial fur.

Cultural Factors

Culture covers everything from religion, beliefs and laws to languages, food, dress, art and traditions.

1) If you're designing a product aimed at a <u>specific target market</u>, you'll have to <u>take into account</u> the <u>views</u> and <u>feelings</u> of people from that <u>particular culture</u>.

2) For example, using <u>transparent fabrics</u> in garments may <u>offend people</u> in some <u>cultures</u>, and in <u>China</u>, <u>black</u> can be seen as <u>bad luck</u>.

REVISION TASK

There's lots to think about when choosing a material...

Environmental, social and ethical factors can be tricky to get your head around, as some issues come under more than one of these categories. Draw a mind map that includes the definitions of each of these three factors and the issues that come under each one. It'll really help them to stick.

Forces and Stresses

Forces and stresses are part of everyday life — stuff is always being stretched, squashed, bent and twisted. Objects are designed to cope with these forces, and in some cases, they can use them to their advantage...

Forces Can Act on Objects in Different Ways

1) Forces can cause objects and materials to break or change shape. Materials are strong if they are good at withstanding a force without breaking (see p.18). Force is measured in newtons (N).

2) Stress is the force per unit area. It's often measured in N/m^2.

3) There are a few different types of forces and stresses that you need to know about:

Tension

1) Tensile forces act to stretch an object and pull it apart: ←□→
   E.g. in a tug-of-war, the rope is under tension caused by people pulling in opposite directions.

2) Trampolines use springs with a high tensile strength — when you land on a trampoline, you stretch the springs. The springs resist being permanently stretched and try to return to their original length — this throws you back up into the air.

3) Suspension bridges use cables with high tensile strength to support the weight of the bridge.

If these things didn't have a high tensile strength, they'd snap as a result of the forces acting on them.

Cables under tension

Bridge support under compression

Compression

1) Compression is the opposite of tension. Compressive forces are directed towards one another, and act to squash and shorten objects: →□←

2) Compression often occurs when a material is supporting a weight above it. Chair legs support the person sitting on the chair and bridge supports hold the weight of the bridge. They're made from materials that have a very high compressive strength.

Shear

1) Shear forces often act in opposite directions but unlike compressive and tensile forces, they aren't aligned with one another. This can cause parts of an object to slide past one another.

2) For example, scissors use shear forces to cut (shear) materials.

3) Industrial cutting machines such as guillotines (see p.66) and die cutters (see p.67) also use shear forces to cut materials.

Forces on the top and bottom surfaces act in opposite directions.

The scissor blades push the paper in opposite directions, causing it to be ripped or cut.

Bench legs push up on the bench.

Bending

1) A bending force is a type of shear force that causes materials to bend.

2) This commonly occurs when a load is applied to a certain part of a material — e.g. a bench has to resist bending forces when someone sits down in a certain place.

Torsion

1) Torsion acts to twist objects and materials. The forces attempt to rotate the different parts of a material in opposite directions.

2) Propeller shafts connect a ship's engine to the propeller, which spins in the water and drives the ship forwards. These shafts have to be able to resist torsion — it allows them to transfer rotary motion without breaking or twisting.

Engine rotates the propeller shaft

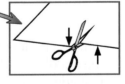

Propeller shaft

Water resists the rotation of the propeller and shaft.

The water applies a rotary force in the opposite direction to the engine — so a twisting force is applied to the shaft.

Forces and Stresses

Materials Can be **Reinforced**, **Stiffened** or Made **More Flexible**

Materials can be <u>reinforced</u> (strengthened) or <u>stiffened</u> (made more rigid) to help <u>resist</u> certain types of <u>forces</u>. They can also be <u>made more flexible</u> to better <u>work with forces</u>. <u>Enhancing</u> materials like this can <u>improve</u> their <u>functionality</u>. Here are a few ways in which materials can be <u>enhanced</u>...

Rigid materials don't deform easily — they're the opposite of flexible materials. Rigidity is different to strength — strong materials don't break easily.

1) <u>Lamination</u> is a process in which one or more <u>layers</u> are <u>added</u> to a material to form a <u>composite</u> (see p.39).

2) It's often used to <u>increase strength</u> and <u>rigidity</u>.

3) For example, <u>foam core board</u> and <u>corrugated card</u> (see p.20) are <u>layered</u> — they're <u>stronger</u> and <u>stiffer</u> than other types of cardboard with a single layer of material.

4) <u>Plywood</u> is made up of <u>layers</u> of wood that are <u>glued</u> together to <u>resist bending forces</u>.

5) <u>Delicate fabrics</u> can also be laminated for extra <u>strength</u>, <u>stiffness</u> and <u>protection</u> (see p.97).

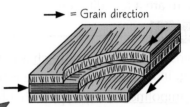

→ = Grain direction

The grain direction in a layer of plywood is at 90° to the layers above and below. This strengthens the weak lines of the grain.

<u>Interfacings</u> are <u>extra layers</u> of fabric <u>stuck</u> or <u>sewn</u> onto the <u>inside</u> of <u>products</u>. They're used in <u>collars</u>, <u>cuffs</u>, <u>pockets</u>, <u>waistbands</u> and <u>button holes</u> — anywhere that needs extra <u>strength</u> and <u>rigidity</u>. Interfacings are used to <u>improve functionality</u> and <u>aesthetics</u>.

<u>Webbing</u> is a <u>fabric</u> (usually cotton, nylon, polyester or polypropylene) that is <u>woven</u> in a way that gives a <u>very high tensile strength</u>. It's used in situations where it will be <u>under tension</u>, e.g. in <u>tow ropes</u>, <u>climbing harnesses</u>, <u>parachutes</u>, <u>seatbelts</u> and <u>backpack straps</u>.

<u>Bending</u> is often used to <u>reinforce</u> and <u>stiffen</u> materials. For example, the middle <u>fluted layer</u> of <u>corrugated card</u> is made from a <u>series of bends</u> that add <u>strength</u> and <u>rigidity</u> — important for packaging <u>heavy loads</u>.

<u>Folding</u> is the <u>bending</u> of a flexible material (e.g. <u>paper</u>, <u>card</u> or <u>fabric</u>) so that the <u>two sides</u> of the bend are <u>flat against each other</u> and a <u>line</u> called a <u>crease</u> or <u>fold</u> occurs <u>between them</u>. <u>Along a fold</u>, a material is <u>made more flexible</u>.

Try folding a thick piece of cardboard... not that easy, right. Now try bending it along the fold you've just made — this should be loads easier.

EXAM TIP

Lamination is a common way of strengthening a material...

You could be asked to describe how forces and stresses act on a material, or how a material can be strengthened or reinforced to withstand forces — make sure you've got some examples in mind. If you get tension and torsion mixed up, remember that to**r**sion involves **r**otation.

Scales of Production

The <u>scale of production</u> is <u>how many</u> of a product you want to <u>make</u>. This decides the <u>type</u> of <u>production</u> that is <u>best suited</u> to the job. There are <u>four</u> main types of production that you need to know about...

One-Off Production Makes **One Product** at a Time

1) One-off production makes <u>bespoke</u> products — <u>every item</u> will be <u>different</u>, in order to meet a <u>customer's</u> exact <u>requirements</u>.

2) For example, <u>wedding dresses</u> can be <u>custom-made</u> to perfectly fit the bride, and <u>made-to-measure furniture</u> is often produced to <u>fit</u> the dimensions of a <u>room</u>.

3) The <u>workforce</u> needs to be <u>highly skilled</u>, especially if the design is quite detailed — so it's an <u>expensive</u> way to make things.

4) This type of production is also very <u>labour-intensive</u> — it takes a lot of <u>time</u> to make each product.

5) <u>Prototypes</u> for a new product are normally produced as a <u>one-off</u>. If the prototype works well, the product might then be manufactured in <u>greater volumes</u>...

Prototypes are full-size working products or systems. They're used to test a product against its design specification (see p.147 for more on this), so they only need to be made in small volumes.

Batch Production Makes a **Certain Number** of Products

1) Batch production is where a <u>specific quantity</u> (a <u>batch</u>) of a <u>product</u> is made. Batches can be <u>repeated</u> as many times as necessary.

2) You do <u>one process</u> (e.g. cutting out) on <u>the whole batch</u>, then do <u>another process</u> (e.g. painting the parts you cut out). So it's <u>quicker</u> than making <u>one-off</u> products over and over again.

3) Batch production can be used to manufacture a <u>load of one product</u> (sofas, say), then a load of something <u>a bit different</u> (e.g. armchairs). <u>Printed circuit boards</u> (PCBs — see p.30) are often batch produced as different electronic products require <u>different PCB designs</u>.

4) <u>Templates</u>, <u>jigs</u> and <u>moulds</u> (see p.53-54) are often used to make sure products <u>within a batch</u> are <u>identical</u>. A design with <u>less detail</u> (than that typically found in <u>one-off production</u>) can help to achieve <u>consistency</u> too.

5) The <u>machinery</u> and <u>workforce</u> used need to be <u>flexible</u>, so they can quickly <u>change</u> to making a batch of a <u>different product</u>.

6) The time <u>between</u> batches, when <u>machines</u> and <u>tools</u> may have to be <u>set up differently</u> or <u>changed</u>, is called <u>down time</u>. This <u>wastes money</u> because you're <u>not making</u> anything you can <u>sell</u>.

Batch production involves making a specific quantity of a product

When revising the differences between the types of production, it's often useful to have a few examples of products that are made by each one. This can help you to remember which is which.

Scales of Production

Mass Production is Making Loads of the Same Product

1) This is the method you'd use to make <u>thousands</u> of <u>identical products</u>, like <u>newspapers</u>, <u>magazines</u> and <u>cars</u>. You'd only use this for a <u>mass-market product</u> — where loads of people want the same thing.

2) The different <u>stages</u> of production are <u>broken down</u> into simple <u>repetitive tasks</u>. Each worker only does a <u>small part</u> of the process, and the product moves further down an <u>assembly line</u> for each stage. Products should have <u>pretty basic designs</u> to make the assembly line simpler.

3) Mass production often uses <u>expensive specialised equipment</u> and <u>CAD/CAM</u> (see p.4).

4) <u>Recruitment</u> is relatively <u>easy</u> — most of your staff <u>don't</u> need to be highly <u>skilled</u>. <u>Robots</u> are increasingly used on assembly lines too (see p.2).

Continuous Production is Non-Stop

1) Continuous production <u>differs</u> from mass production in that it <u>runs all the time</u>, without interruption, <u>24 hours a day</u>. That's because it would be too <u>expensive</u> to keep <u>stopping and restarting</u> the process.

2) It's pretty much entirely <u>automated</u> — very <u>few workers</u> are needed (thankfully)...

Both mass production and continuous production can keep manufacturing costs fairly low — this is partly because the initial cost of equipment is spread over a large number of products.

3) The equipment is built to make <u>huge amounts</u> of only <u>one thing</u>. This means it's expensive but it can be designed to be <u>very efficient</u> — so the <u>cost per item</u> is <u>cheap</u>.

4) Continuous production is used to make products such as <u>aluminium foil</u> and <u>chemicals</u>.

Continuous production — feels more like continuous revision...

So, products can be made as a unique one-off, in a group called a batch, in really massive numbers or in a continuous, never-ending, keep-going-until-the-end-of-time way. Make sure you know the advantages and disadvantages of each type of production — it could be easy marks in the exam.

Quality Control

You should always <u>check</u> a product <u>works</u> and <u>looks right</u> before <u>selling</u> it — this is known as <u>quality control</u>.

Quality Control Checks the Quality of Manufacturing

1) <u>Quality control</u> ensures products are <u>manufactured</u> to a <u>high enough standard</u>. Checks should be made at <u>every stage</u> of the <u>manufacturing process</u>.

2) This involves <u>testing</u> materials, products or components to check they've been made to a <u>high enough standard</u> and they <u>meet</u> the <u>manufacturing specification</u>. E.g. a musical greetings card might be tested to make sure all the images are <u>printed clearly</u>, the edges are <u>cut accurately</u>, the card is <u>folded</u> in the <u>right place</u>, the text is <u>straight</u>, it plays the <u>right tune</u>, etc.

3) Usually, only a <u>sample</u> of the materials, components or products are <u>tested</u>, as it would take <u>too much time</u> to check them <u>all</u>.

Manufacturing specifications (see p.145-146) include manufacturing instructions and exact details of what the product should be like.

Tolerance is the Required Accuracy of a Measurement

1) It's important that the <u>dimensions</u> (size) of components are <u>accurate</u>. If they aren't, the parts won't <u>fit together</u> properly when the product is assembled.

2) Just <u>how accurate</u> they need to be is <u>specified</u> by the <u>tolerance</u>. This is the <u>margin of error</u> that is considered to be <u>acceptable</u>, e.g. small enough <u>not</u> to <u>affect</u> the product's <u>functionality</u>.

3) The tolerance is normally given as an <u>upper</u> (+) and <u>lower</u> (–) <u>limit</u> for the measurement. For example, a <u>20 mm</u> measurement with a tolerance of ±0.5 mm has a lower limit of <u>19.5 mm</u> and an upper limit of <u>20.5 mm</u>.

Tolerance can be applied to loads of different types of measurement — not just size. For example, measurements of physical properties (e.g. weight), chemical properties (e.g. the composition of an alloy) or electrical properties (e.g. the resistance of a resistor, see p.114).

> **EXAMPLE:** A shelf is designed to be 200 mm wide by 720 mm long, with a tolerance of ± 5.0 mm. Would a 203 mm by 728 mm shelf be within this tolerance?
>
> The width should be 200 ± 5.0 mm — between 195 mm and 205 mm, which it is.
> The length should be between 715 mm and 725 mm, which it isn't, so it's not within the tolerance.

4) Tolerances should be included on <u>working drawings</u> (in the <u>manufacturing specification</u>) to show the <u>limits</u> within which the product should be manufactured.

5) Components must be <u>within the tolerance</u> to <u>pass</u> through any <u>size checks</u> in quality control.

There are Many Different Quality Control Tests

Go/No Go Fixtures Check Dimensions are Within Tolerance

1) Go/no go fixtures are <u>limit gauges</u> — they <u>check</u> to see whether the size of a part is <u>within its tolerance</u>.

2) They're usually <u>double-ended</u> — one end is machined to the <u>lower limit</u> and the other end to the <u>upper limit</u> of tolerance.

3) These checks <u>don't take very long</u> and are <u>much quicker</u> than <u>measuring</u> the <u>actual dimensions</u> of a component, e.g. using a ruler or micrometer.

4) They're often used in the quality control of <u>timber-based</u> components.

If the component <u>fits through</u> the <u>upper limit</u> (go) but <u>not</u> the <u>lower one</u> (no go), it's <u>within the tolerance</u>.

Registration Marks Check Printing Plates are Aligned

1) A <u>colour registration mark</u> normally appears as a <u>cross shape</u> — it is used by manufacturers to <u>check</u> the <u>quality</u> of <u>colour printing</u> onto <u>paper</u> and <u>board</u>.

2) They're used to make sure the <u>printing plates</u> are <u>aligned</u> in the right position.

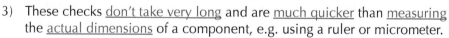

If the plates <u>are</u> all in the right places you get a single, <u>clear image</u>...

... and if not, the <u>registration mark</u> will be printed a bit <u>fuzzy</u>.

Colour printers use four printing plates — cyan, magenta, yellow and black (see p.69).

Quality Control

Prints Should be Checked Against the Original

1) To check repeating prints (such as stripes and chequers) are being printed correctly onto fabrics, manufacturers will often compare prints to an original sample print. This can be done pretty well by eye.

2) It's much quicker and easier to do this than try to measure the dimensions of each part of the repeating pattern one at a time.

There are Ways to Achieve Consistency During Manufacture

Depth Stops Control the Depth of Drilling and Cutting

1) Depth stops are long rods that are clamped close to the drill bit of some drills.

2) They allow you to drill a hole to an exact depth in whatever material you're drilling. Once this depth has been reached, the depth stop will come into contact with the material and will prevent you from drilling any deeper.

3) The drill depth can be adjusted by clamping the depth stop in a different place.

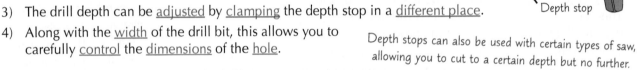

Depth stop

4) Along with the width of the drill bit, this allows you to carefully control the dimensions of the hole.

Depth stops can also be used with certain types of saw, allowing you to cut to a certain depth but no further.

Laser Cutters Need to be Programmed Correctly

1) To cut a material accurately, a laser cutter needs to be programmed with the right information. This includes:

There's more on laser cutters on page 5.

- The dimensions of the component to be cut — so it knows exactly where to cut.
- The correct power settings and feed rate (the speed that the laser moves over and cuts the material). These values depend on the type of material to be cut and its thickness.

2) If programmed correctly the laser can cut to a tiny tolerance (with high accuracy and precision), so the component should end up as you want it.

3) A sample still needs to be checked though to make sure the machine hasn't malfunctioned or been incorrectly programmed.

PCBs Should be Made with the Same Exposure Times

1) Printed circuit boards (PCBs) can be manufactured through photo-etching.

2) In this process, the PCB is exposed to UV light, a developer solution and an etching solution — this removes unwanted copper from the board, leaving behind the copper needed to form the tracks.

3) Exposure times have to be carefully selected. For example, exposing the PCB for too long will cause the copper tracks to be removed, but too little time will mean unwanted copper remains on the board.

Photo-etching is covered on page 118.

4) Once chosen, exposure times should be kept the same for every PCB that's produced. This helps to make sure that the PCBs are all of a consistent quality when output.

All manufacturers should have some sort of quality control in place...

Quality control is a really important part of the manufacturing process — it's important to check the quality of what you're making. It can take a while to get your head around things like tolerance, so take your time and make sure you're happy with everything on the last couple of pages before turning over.

Warm-Up and Worked Exam Questions

Alright, time to put all that revision into practice and see how much you've understood in this section...

Warm-Up Questions

1) What is meant by the term 'social responsibility'?
2) What type of force would be used to twist the lid off a jar?
3) Explain what is meant by 'down time' in batch production.
4) A company wants to manufacture 100 000 identical nappies.
   a) What type of production would be best suited for this?
   b) What is the advantage of this type of production when it comes to hiring workers?
5) a) What is the purpose of quality control?
   b) Explain why only a sample of the manufactured items are used in quality control.

Worked Exam Questions

1 Sheets of glass are usually produced by continuous production.

a) What is meant by the term 'continuous production'?

A type of production that goes for 24 hours a day without stopping at any point.

[1 mark]

b) Give **one** advantage of using continuous production.

You could have also said that the cost per item is low.

The processes used in continuous production can be made very efficient.

[1 mark]

2 **Figure 1** shows a man in a hammock.

a) Force A acts on the ropes that attach the hammock to the beam above. Force B acts on the hammock due to the weight of the man. Name these **two types** of force.

Force A: *tension*

Force B: *bending*

[2 marks]

Figure 1

b) Some hammock straps are made from webbing.

i) What is webbing?

Webbing is a woven fabric with a high tensile strength.

[2 marks]

ii) Suggest why webbing is sometimes used to make hammock straps.

Webbing has a high tensile strength, so it won't break easily when the straps are under tension during use.

[1 mark]

Section Three — More about Materials

Exam Questions

1 An outdoors company wants to sell a folding table to be used by campers.

a) The designers of the table are trying to select a material for the legs and table top.
Give **two** functional considerations that they should bear in mind.

1. ....

....

2. ....

....
[2 marks]

b) The company wants the table to be manufactured using materials in stock forms and sizes
as they are widely available. Give **one** benefit of using widely available materials.

....

....
[1 mark]

c) The company will buy the materials in bulk as they intend to mass produce the table.
How might bulk buying the materials affect the price at which the table is sold at?
Explain your answer.

....

....

....
[2 marks]

2 There are many ways of ensuring consistency and accuracy between products
during manufacture.

a) A depth stop can be used to ensure holes are drilled to an exact depth.
Describe how depth stops ensure consistency when drilling.

....

....
[2 marks]

b) Laser cutters need to have the correct settings selected for the laser to cut out
a material accurately. Give **two** settings which must be correctly selected.

1. ....

2. ....
[2 marks]

c) Photo-etching is a process that is used to manufacture printed circuit boards (PCBs).
Give **one** way of ensuring PCBs are made consistently during photo-etching.

....
[1 mark]

Production Aids

Production aids do exactly what they say on the tin — they're tools and techniques used to aid production. They often speed up or simplify a process, or are used to help control accuracy...

All **Points** Can be Defined **Relative** to a **Reference Point**

1) A reference point is a point where measurements are made from — this could be the corner of an object. You can also have reference lines (e.g. the x-axis, or an object's edge) and reference surfaces that you measure from.

2) Using reference points, lines and surfaces helps to control accuracy, e.g.

Imagine you want to make a coat rack with a coat hook every 10 cm. You wouldn't measure each hook 10 cm from the previous one, as if the previous hook is slightly off, the inaccuracy would be carried over. Instead, you'd mark out where each hook will go by measuring each one from a single reference point, e.g. at one end of the coat rack — this is quicker and reduces measurement error.

3) When x,y,z coordinates are used, the reference point is often the datum — the point where x, y and z meet (0,0,0).

4) Reference points are really important in aligning CAM machines with the materials they're operating on. CAM machines also use x,y,z coordinates to guide the tools being used (see p.4).

5) Reference points are useful when scaling the size of something too, e.g. a drawing:

- First, choose a reference point, e.g. one corner of the drawing.
- Then, figure out how far away each point on your drawing is from the reference point. This is basically their (x,y) coordinates if the reference point is the datum, i.e. (0,0).
- Next, multiply these coordinates by a scale factor — for example, if you wanted to make the drawing twice as big, you'd multiply by two, e.g. (0,4) would become (0,8).
- Finally plot your new coordinates and ta-da, you've scaled your drawing.

Scaling can be really useful. For example, you might want to scale a pattern (see next page) to produce a piece of clothing in a different size.

Templates are Used for **Drawing** and **Cutting Round**

1) Templates are very easy to make and simple to use.

2) You can draw round or cut round them to produce a shape that is identical to the template.

3) This is much quicker and easier than measuring out the dimensions of the shape each time.

4) As long as your template is accurate the batch of components you make with it should all be accurate too — using a template makes the products consistent.

5) Templates need to be strong and hard-wearing — so that they can be used repetitively without getting damaged or worn.

6) They're useful in quality control too — you can check your components against the template to make sure they're the same.

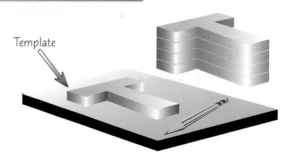
Template

If you're cutting lots of shapes from a sheet of material, you need to plan how you want to lay the template out — some layouts will result in less wastage of material (e.g. see p.150).

Production Aids

Patterns are Used with Textiles and for Casting

1) In textiles, patterns are <u>templates</u> that are used to <u>cut out pieces</u> of <u>fabric</u> — these <u>pieces</u> are then <u>sewn together</u> to make a <u>textiles product</u>.

2) <u>Patterns</u> are normally made from <u>tissue paper</u> — they're <u>pinned</u> to a fabric so that you can <u>cut round</u> them.

3) In <u>industry</u>, <u>CAM cutting machines</u> are used to <u>cut</u> the fabric. The <u>patterns</u> are drawn out using <u>CAD</u>. The <u>coordinates</u> of the <u>pattern's outline</u> are then used to tell the machine <u>how to cut</u> the pieces of fabric.

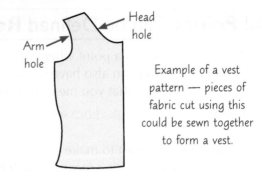

Head hole

Arm hole

Example of a vest pattern — pieces of fabric cut using this could be sewn together to form a vest.

1) <u>Casting</u> involves forming a material into a <u>particular shape</u> — see p.87-88. This shape is determined by a <u>pattern</u> — which is an <u>exact replica</u> of the object you want to form.

2) Patterns can be made from materials such as <u>resin</u>, <u>wood</u> and <u>metal</u>.

3) Patterns can be <u>used many times</u> to make moulds that are used in the casting process and make products that are a <u>consistent shape</u>.

Example of a metal part that has been cast using a pattern.

Jigs Guide Tools to Simplify and Speed Up Production

1) A jig <u>guides</u> the <u>tools</u> that are working on a component, or makes sure that the <u>workpiece</u> is <u>positioned</u> in the <u>right place</u> for machining.

2) <u>Jigs</u> come in many <u>different shapes and sizes</u> and can be <u>specifically</u> made for a particular job.

3) They're designed to <u>speed up production</u> and <u>simplify</u> the <u>making</u> process.

Although jigs are really useful, they can take a long time to make. So it's not worthwhile using one if you're only going to make a few products.

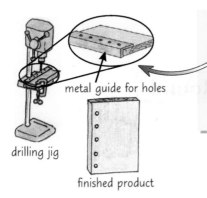

metal guide for holes

drilling jig

finished product

Using a <u>drilling jig</u> means you <u>don't</u> have to <u>mark out</u> exactly <u>where</u> you want <u>to drill</u> — this <u>saves time</u> and <u>effort</u>. It <u>guides</u> the drill, which helps to cut down on <u>errors</u>, and makes sure that components are <u>consistent</u>.

A <u>dovetail jig</u> allows complex, <u>dovetail joints</u> to be cut <u>quickly</u>, <u>easily</u>, and with <u>minimal measuring</u> and marking out.

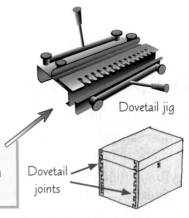

Dovetail jig

Dovetail joints

No — there's nothing on these pages about Scottish dancing...

REVISION TASK

Production aids make a manufacturer's life much easier. Sadly, you'll need to write a bit more than that in the exam. List the production aids covered on these pages — for each one, write down how they can be used to make production easier. And don't cheat — you'll only be cheating yourself.

Production of Materials

New materials need to be extracted from a raw material source and processed into a useful form.

Paper and Board are Made from Cellulose Fibres

1) Trees go through a series of processes before they become paper or board:

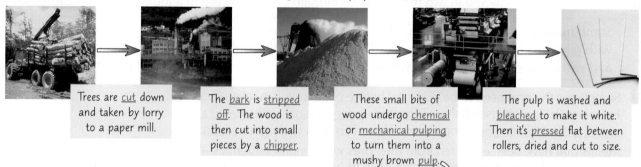

Trees are cut down and taken by lorry to a paper mill.

The bark is stripped off. The wood is then cut into small pieces by a chipper.

These small bits of wood undergo chemical or mechanical pulping to turn them into a mushy brown pulp.

The pulp is washed and bleached to make it white. Then it's pressed flat between rollers, dried and cut to size.

Some types of paper and board aren't bleached and remain a brown colour.

> Pulping is a very important stage — it converts the wood into individual cellulose fibres (pulp). Mechanical pulping involves grinding down the wood to separate out the fibres. Chemical pulping involves heating the wood with chemicals — this dissolves other parts of the wood, leaving just the fibres behind.

2) The cellulose fibres used for paper and board can also come from other plants like grasses. The process is similar to the conversion of wood to paper, although no debarking or chipping is required.

3) Certain types of board may have an extra stage in production. For example, corrugated cardboard will need its fluted middle layer sticking to the outer layers to form a laminate.

Wood is also Cut Down for Timber

1) Trees, e.g. from a plantation or a forest, are felled (cut down).

2) The bark is removed, and the trunk is sawn up. Which way it's sawn affects how the planks will look, and how much they're likely to bend or warp (twist).

3) The wood is then seasoned by drying it. Wood can be 'air-dried' or 'kiln-dried'...

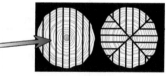

Slab sawn   Quarter sawn

> • Air-drying involves stacking wood up outside or in sheds and allowing air to circulate around it.
>
> • Kiln-drying is similar to air-drying but it's done in a heated chamber with fans to circulate the air. This makes it loads quicker, but the conditions have to be carefully controlled to avoid the wood drying too fast (which can damage it).

4) Seasoned wood can be converted into lots of different useful forms, e.g. planed square edge wood, mouldings, etc. There's more about the forms wood is available in on p.77.

5) Wood can also be processed into manufactured timbers (see p.27):

Seasoning makes the wood stronger and less likely to rot (see p.76).

> • Medium density fibreboard, MDF uses wood that has been processed into cellulose fibres, in a similar way to paper and board production (e.g. debarked, chipped, and pulped).
>
> • Chipboard uses a mixture of dried wood chips, shavings and sawdust — this is often material that has been discarded as waste from another process.
>
> Glue (e.g. urea formaldehyde) is added to these mixtures. They're heated, pressed into sheets and left to dry.

> Plywood uses wood that has been softened, either through soaking it in hot water or steaming it. A thin sheet is peeled from the softened wood, cut to suitable size and dried. The cut wood can then be arranged into stacks of three or more layers. Each layer has a grain direction that is at 90° to the layer above and below (see p.46). Glue is added between each layer and the sheets are heated and pressed.

Once they're dry, manufactured boards can be trimmed and sanded to give them a smooth finish.

Production of Materials

Metals Come from Rocks in the Ground

1) Metal is <u>mined</u> from the <u>ground</u> as a <u>metal ore</u> — a rock with enough metal locked up in it to make it <u>profitable</u> for the metal to be <u>extracted</u> (separated) from the rest of the ore.

2) <u>Extraction</u> can be achieved in <u>two</u> main ways — through <u>heating</u> the ore in a <u>furnace</u> or by <u>electrolysis</u>:

> Some metal ores can be <u>crushed</u> and then <u>heated</u> in a <u>furnace</u> with <u>other materials</u> such as <u>charcoal</u>. The metal <u>separates</u> out and will often <u>sink</u> to the bottom where it can be <u>tapped off</u>.

> <u>Electrolysis</u> uses <u>electricity</u> to extract a metal from its ore.

3) Sometimes the ores need to be <u>prepared</u> before extraction can take place. For example, <u>bauxite</u> (see below) is <u>heated under pressure</u> with <u>chemicals</u> to <u>separate aluminium oxide</u> from the mixture. This then undergoes <u>electrolysis</u>.

4) Metals that have been extracted from their ores will <u>contain impurities</u> — <u>small amounts</u> of other <u>substances</u>. These impurities will often need to be <u>removed</u> before the metal can be used in a product. This process is known as <u>refining</u>, and there are several ways of achieving this (see below).

5) The refined, <u>molten</u> metal is usually <u>cast</u> — this process involves pouring the metal into a <u>mould</u> and allowing it to <u>cool</u> and <u>solidify</u> into the <u>required shape</u>.

6) Here's a bit more on the processes involved in the <u>production</u> of a few different metals...

Metal	Common ore	Extraction method	Refining method
Iron	Haematite	Furnace	Main impurity is sand — removed by adding limestone to the furnace during extraction.
Aluminium	Bauxite	Electrolysis	Gases can be pumped through molten aluminium. This causes impurities to rise to the surface.
Zinc	Sphalerite	Furnace or electrolysis	Zinc is heated and then cooled. This separates it from impurities with different boiling points.
Copper	Chalcopyrite	Furnace	Electrolysis can be used to separate the copper from any impurities.
Tin	Cassiterite	Furnace	Heating the tin can cause some impurities to react and form a substance that can easily be removed. It can also be used to separate impurities based on melting points. Electrolysis can be used to refine the tin too.

Most Plastics are Made from Oil

Most plastics are made from <u>crude oil</u>, although some can be made from substances found in <u>plants</u>.

1) <u>Crude oil</u> is <u>extracted</u> from the <u>ground</u> (on land or at sea) and taken to a <u>refinery</u>.

2) Here, it is <u>heated</u> using a process called <u>fractional distillation</u>. This <u>separates</u> it into different chemicals called 'fractions'.

3) <u>Some</u> of the fractions can be <u>linked together</u> to make <u>polymers</u> (plastics):

> • The chemicals that get joined together are called <u>monomers</u>.
> • The monomers join up to make <u>chains</u> called <u>polymers</u>.
> • So the process is called <u>polymerisation</u>.

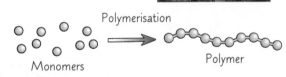

Polymerisation

Monomers → Polymer

4) <u>Other fractions</u> may need to be <u>broken down</u> into <u>smaller molecules</u>. This is done by <u>heating</u> these fractions in a process called <u>cracking</u>. These chemicals can then be <u>polymerised</u>.

5) The plastics can then be <u>used</u>. For example, they are often <u>cut up</u> into <u>small pieces</u> and <u>moulded</u> into the <u>right shape</u> (see pages 88-90 for more on how plastics can be moulded).

Remember — many monomers join together to make one polymer...

This stuff can be tricky — make sure you know the steps involved in getting each material into a useful form.

More on the Production of Materials

Fabrics are made up of fibres. There's a huge range of fibres that are used in the textiles industry.

There are **Many Types** of **Fibres**

1) Fibres are tiny 'hairs' that are spun into yarns — the threads that are woven or knitted into fabrics.

2) Fibres come in either short lengths (staple fibres), or longer lengths (filaments). Filaments can be cut up into staple lengths if required.

3) Fibres fall into three main categories — natural, regenerated and synthetic fibres...

Natural Fibres are **Harvested** from **Natural Sources**

1) Natural fibres are fibres obtained from natural sources (e.g. plants and animals).

2) First, they're harvested and processed, for example...

WOOL is from sheep's fleece

The sheep is sheared — the best wool comes from the sheep's shoulders and sides.	Cleaning and scouring (washing with harsh chemicals) removes grease and dried sweat.	The fibres are then combed using wire rollers — this is called carding.

COTTON is from the seed pods of the cotton plant

Before harvesting, the plants are treated with chemicals to make the leaves fall off.	Fibres are cleaned to remove dirt.	Seeds are removed.	Carding (see above).

SILK comes from the cocoon made by silk worms

cocoon

The cocoon protects the worm whilst it turns into a moth.	The cocoon is one long piece (a filament) of silk fibre. It's stuck together with a hard gum (called sericin).	The gum is softened by soaking in warm water. The filament is then unwound (this is called reeling).

3) The fibres can then be spun into yarn. These are then made into fabric — see p.25.

Regenerated Fibres are **Chemically Treated Natural** Materials

1) Regenerated fibres are made from natural materials — usually cellulose from wood pulp. However, these materials are chemically treated to produce fibres.

2) Different fibres are made by using different chemicals — e.g:

> Viscose fibres are made by dissolving cellulose in sodium hydroxide solution. The liquid is forced through tiny holes, and hardened to form filament fibres. These filaments are stretched into a yarn which is wound onto spools, or chopped into staple lengths.

You can think of regenerated fibres as somewhere between natural and synthetic fibres — they are semi man-made.

Synthetic Fibres are Made from **Polymers**

Synthetic fibres are man-made fibres. They're made from polymers — long chains of monomers that come mainly from oil. For example, polyester, LYCRA® and nylon are produced from crude oil...

Different polymers will make different synthetic fibres.

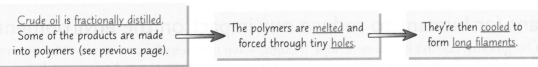

Crude oil is fractionally distilled. Some of the products are made into polymers (see previous page).	The polymers are melted and forced through tiny holes.	They're then cooled to form long filaments.

The filaments are made into yarn. This is wound onto spools, or chopped into staple lengths to be spun.

More on the Production of Materials

Material **Production** Can **Impact** the **Environment**

1) The conversion of raw materials into useable forms can have a negative effect on the environment — it can contribute to pollution, the destruction of natural habitats and climate change. (There's more on this on p.8 and p.44.)

2) Just obtaining the raw materials causes environmental problems. For example:

Obtaining wood for timber, paper and board can cause deforestation. This is where large areas of forest are cut down, and trees aren't planted to replace the old ones. This destroys the forest habitat, which has a negative effect on the animals and plants that live there.

Sustainably managed forests (where trees are replanted) are a much more environmentally friendly source of wood.

Drilling for oil is the first step in the production of polymers and synthetic fabrics. Oil can be drilled both on land and out at sea. Drilling can release toxic chemicals into the atmosphere. Sludgy waste material or oil can also leak and pollute surrounding land or marine habitats — this can be harmful to the wildlife living there. On land, habitats need to be cleared to make room for the drill sites.

Mining metals uses a lot of energy from fossil fuels — this causes air pollution and contributes to global warming. Habitats are destroyed when large areas of land are cleared for mines. The surrounding habitats can be affected too — chemicals (used in mining) and waste rock can pollute nearby water, which can be harmful to wildlife.

Oil and metals are both examples of finite resources — once they've been used up there won't be any more (see page 6 for more about this).

Pesticides are chemicals used to kill weeds and insects that harm the crops. Fertilisers are chemicals that help crops to grow.

Farming natural fibres (e.g. cotton) often uses artificial fertilisers and pesticides to help increase the amount of natural fibre produced — these can pollute rivers and harm wildlife. Farming also needs lots of land — getting it may involve deforestation or clearing habitats.

It's important for humans to reduce our impact on the environment...

This all sounds a bit gloomy, doesn't it. There is hope though — we can help the environment by reusing and recycling the materials we've already produced. That way, we don't have to produce as much new stuff, reducing the damage to the environment. That's the theory, at least. It's often harder to put into practice...

Warm-Up and Worked Exam Questions

It's that time again — first see if you've understood the basics by tackling these warm-up questions.
After that, follow the worked exam questions and see if you can do the questions on the next page yourself.

Warm-Up Questions

1) What is commonly used as the reference point when working with x,y,z coordinates?
2) Describe how a template is used as a production aid to make identical shapes from a material.
3) What is the name of the raw material that metals are extracted from?

Worked Exam Questions

1   A company is producing a batch of shelving units. The shelving units need side panels that have pre-drilled holes for the shelf supports. They have decided to use a drilling jig to make the holes.

a) Describe what a jig does.

A jig helps to guide tools when working on a component and makes sure
that the workpiece is positioned in the right place when working on it.

[1 mark]

b) Give **two** advantages of using a drilling jig instead of marking out and drilling the holes by hand.

You could have also mentioned that it makes the pre-drilled holes consistent for all of the shelving units in the batch.

1. It can help to reduce human error.

2. It can save time and effort.

[2 marks]

2   Paper and card are made from cellulose fibres, which can come from wood chips.

a) Explain how wood chips can be converted into cellulose fibres.

The wood chips are heated with chemicals. This dissolves non-fibrous parts of the
wood, leaving only the cellulose fibres behind.

This answer describes chemical pulping. You could have talked about mechanical pulping instead though.

[2 marks]

b) Describe the process used to convert these fibres to white paper.

The pulp containing the cellulose fibres is washed and bleached to make it white.
It is then pressed flat between rollers, before being dried and cut to size.

[2 marks]

c) Name **one** other raw material that can be used to make paper and board.

Grasses

[1 mark]

Exam Question

1   Cotton and polyester fibres are commonly used to make clothes.

   a)   What are the raw materials used to make cotton and polyester yarns?

   Cotton: ....

   Polyester: ....

   [2 marks]

   b)   Explain how obtaining these raw materials can negatively impact the environment.
   Give **two** ways for each material.

   i)   Cotton

   1.   ....

   ....

   2.   ....

   ....

   [2 marks]

   ii)   Polyester

   1.   ....

   ....

   2.   ....

   ....

   [2 marks]

   c)   Choose **one** of these raw materials.
   Use sketches and/or notes to give a detailed description of how it is converted to yarn.

   Material: ....

   [4 marks]

Revision Questions for Section Three

Well, that's Section Three all wrapped up — time to see how much you can remember.
- Try these questions and tick off each one when you get it right.
- When you've done all the questions for a topic and are completely happy with it, tick off the topic.

Selecting Materials (p.43-44) ☑

1) Give three factors that you should consider when selecting a material to use in a product. ☑
2) Suggest two disadvantages of using a material that is not widely available. ☑
3) Isaac is hand-crafting a one-off chair made from expensive mahogany.
   Suggest two reasons why Isaac needs to sell it at a high price. ☑

Forces and Stresses (p.45-46) ☑

4) Which type of force acts to squash or shorten an object? ☑
5) The photo on the right shows someone abseiling down a wall.
   What type of force is acting on the climbing rope? ☑
6) Name one type of industrial cutting machine that uses shear to cut materials. ☑
7) Describe how you could apply torsion to a rod. ☑

Scales of Production (p.47-48) ☑

8) Put one-off, batch and mass production in order of scale (from smallest to largest). ☑
9) What type of production is normally used to make a product prototype? ☑
10) How is batch production carried out? ☑
11) Suggest an example of a product that is likely to be mass produced. ☑
12) Give one reason why continuous production runs non-stop. ☑

Quality Control and Production Aids (p.49-50 & p.53-54) ☑

13) A board needs to be cut with dimensions of 135 × 180 mm with a tolerance of ± 3 mm.
    Which of the following falls within the stated tolerance?
    A: 38.1 × 177.5 mm   **B:** 134.4 × 183.2 mm   **C:** 133.9 × 181.8 mm   **D:** 138.0 × 176.7 mm ☑
14) What is a go/no go fixture used for? ☑
15) Describe what a registration mark is used for. ☑
16) a) What is a reference point?
    b) Give two ways in which reference points aid production. ☑
17) What are patterns for textiles products normally made from?
    A: fabric   **B:** tissue paper   **C:** board   **D:** tracing paper ☑
18) Adam is making a one-off product. Explain why using a jig won't save him time overall. ☑

Production of Materials (p.55-58) ☑

19) a) Which raw material are plastics usually made from?
    b) What is fractional distillation? ☑
20) a) Which of the following is an example of a regenerated fibre?
       A: cotton   **B:** silk   **C:** viscose   **D:** polyester
    b) Name two synthetic fibres. ☑
21) Give two ways in which mining metals can harm the environment. ☑

Properties of Paper and Board

Each type of paper and board has <u>properties</u> that make it <u>more suited</u> for certain tasks. Let's have a look at some of the <u>physical properties</u> that determine how they're used in different products...

Different **Properties** are **Useful** for Different Things

1) <u>Different sorts</u> of paper and board have <u>pros</u> and <u>cons</u> — you need to be ready to <u>weigh them up</u>.

Flexibility and rigidity

To make 3D products, you need paper and board that can be <u>bent</u> or <u>folded without breaking</u> but is <u>rigid</u> enough to <u>keep its shape</u> (if it's used as a box, say). <u>Corrugated board</u> is a good option when you need quite a strong, stiff material.

Toxicity and sustainability

1) Recycled paper and board may contain <u>toxic chemicals</u> which mean they're not suitable for use as food packaging.

2) <u>Laminated paper and board</u> (see p.70) can be <u>hard to recycle</u> because it's hard to separate the paper/board from the other materials.

Strength and weight

1) Some materials (like <u>corrugated card</u>) can withstand a fair amount of <u>force</u> without breaking — good for <u>heavy duty packaging</u> or products that will be <u>handled</u> a lot.

2) But stronger materials are usually <u>bulkier</u> which can impact <u>transport costs</u> — the more bulky they are, the more they cost to transport.

Cost and quality

1) Expensive paper and board makes a product feel <u>high-quality</u>.

2) But it is only worth using for <u>luxury products</u> or something that has to <u>last a long time</u>.

Finish

1) <u>Solid white board</u> has an excellent surface for <u>printing</u> on.

2) A <u>cheaper alternative</u> is to use <u>recycled board</u> laminated with <u>high quality paper</u>.

3) <u>Plain recycled board</u> can be used for something that doesn't need a good finish.

2) When <u>choosing</u> a material, you need to <u>consider its use</u>. Here are some examples:

gsm stands for grams per square metre.

Card based food packaging

1) Cardboard food packaging needs a <u>combination of properties</u>. It needs to be <u>printable</u>, so that the customer can see the advertising and nutritional information for what they are buying. But it often also has to be <u>waterproof</u> and <u>airtight</u>, to prevent the food inside from going bad.

2) A combination of <u>aluminium foil</u> and <u>board</u> can be used to package food, e.g. soup cartons. It keeps <u>flavours in</u> and <u>air out</u>. You can also print graphics onto the paper. Duplex board (look back to p.20) is often used for food packaging because it is <u>strong</u> and <u>easy to print on</u>.

3) <u>Pizza boxes</u> are made of corrugated cardboard that is <u>strong</u> enough to withstand other boxes being <u>stacked on top of it</u>. It's <u>thick</u>, so it's <u>good at retaining heat</u>, which keeps the pizza warm.

Flyers and leaflets

1) Flyers and leaflets need to be <u>produced cheaply</u> and often in <u>large quantities</u>. Usually the <u>quality of print</u> doesn't need to be high as it doesn't need to last for a long time. For these reasons, they're often printed on <u>cheap</u>, <u>low-weight</u> (low gsm) paper.

2) Some leaflets for <u>classier products</u> might be printed on <u>heavier, higher quality paper</u> because it suits a higher quality product.

3) <u>Biodegradable paper</u> would work well, as the leaflets are often <u>dropped on the street</u> or <u>thrown away</u>.

4) <u>Unbleached paper</u> may also be a good option as it is more <u>environmentally friendly</u>.

If you laminate paper by adding a layer of another material, you get a composite (see p.39) with different properties.

Properties of Paper and Board

Properties of Paper and Board can be **Changed**

1) Paper and board can be <u>modified</u> for a specific purpose. <u>Chemicals</u> called additives can be added <u>during manufacture</u>, or can be used to coat paper and board to give it desirable properties.

2) For example, <u>chemical additives</u> can increase <u>paper strength</u>, enhance <u>brightness</u> or add <u>colour</u>.

3) Another example is to <u>prevent moisture transfer</u>, i.e. moisture <u>passing through</u> the paper or board.

Examples

1) <u>Kraft paper</u> undergoes a chemical treatment that makes it <u>strong and tear-resistant</u>. It's also a <u>vapour barrier</u>, so can be used on food packaging for things like flour, which needs to <u>stay dry</u>.

2) <u>Baking parchment</u> is usually treated with sulphuric acid to give it its <u>non-stick properties</u>.

3) <u>Greaseproof paper</u> is treated to change the qualities of its fibres — making it <u>less absorbent</u>.

4) <u>White board</u> is a <u>strong</u>, <u>thick</u> board that is <u>good for printing on</u>.
It's the only type of board recommended for <u>direct contact</u> with food.
The board is <u>bleached</u> and then coated with <u>wax</u> (or laminated with polyethylene).

5) The pulp used to make <u>paper towels</u> is treated with <u>different resins</u> during manufacture to increase its <u>wet strength</u>. This makes sure that the towels <u>don't break</u> when they become <u>wet</u>.

Paper and Board Come in Different **Shapes** and **Sizes**

1) There are <u>various types</u> of paper and board available (take a look back to p.20).

2) Paper and board can be bought by the <u>sheet</u> and in <u>rolls</u> — these are known as <u>stock forms</u> (also see p.77).

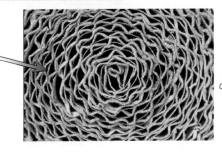

A roll of corrugated board

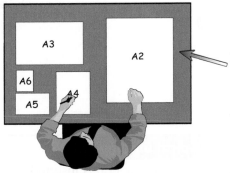

3) Sheets of paper and board are sold in <u>standard sizes</u>. Sizes go <u>from A0</u> (which has an area of 1 m²) to <u>A1</u>, <u>A2</u>, and so on — halving in size (area) each time.

4) Many other sizes are also available — <u>A4</u> is <u>half</u> the size of <u>A3</u>, <u>A5</u> is <u>half</u> the size of <u>A4</u>, <u>A6</u> is <u>half</u> the size of <u>A5</u>, etc.

5) As the size gets smaller the number <u>increases</u>.

6) The most common paper sizes used in schools are <u>A4</u> and <u>A3</u>.
The <u>width</u> of A3 paper is the <u>length</u> of A4. The <u>length</u> of A3 paper is <u>double the width</u> of A4.

7) Paper and board are also available in different <u>weights</u> (see p.20), <u>thicknesses</u> and <u>colours</u>.

8) You can choose the <u>ply</u> of your material too. Ply means how many <u>layers</u> the material is made out of — 1-ply is one layer, 2-ply is two layers and so on.

Different types of paper and board have different properties...

When you're picking a type of paper or board for a project, you need to think about what you need it to do and select the right properties. Don't forget that you can change the properties too...

Standard Components

Standard components are parts that come <u>ready-made</u> for use. Here's two pages filled with them...

Standard Components are Pre-Manufactured Parts

1) Standard components are common fixings and parts that <u>manufacturers buy</u> instead of manufacturing them <u>themselves</u>, e.g. screws, rivets and buttons.

2) Standard components are <u>mass produced</u> so they're available at <u>low cost</u> to manufacturers.

3) Using standard components saves <u>time</u> during manufacture — so it is more <u>efficient</u>.

4) <u>Specialist machinery</u> and <u>extra materials</u> aren't needed so it also saves <u>money</u>.

5) So, when <u>designers</u> are working on a new product, they need to think about which standard components they could use. When using CAD, designers can <u>select</u> standard components to use in their designs.

Lots of Standard Components are Used with Paper and Board

<u>Velcro®</u> pads are self-adhesive pieces of the <u>famous two-part hook and loop system</u>. They've got loads of uses — they're particularly good for displays.

<u>Treasury tags</u> hold stuff together <u>loosely</u>. They're really cheap.

Drawing pins (also known as <u>thumb tacks</u> or <u>mapping pins</u>) are useful for sticking paper and card to <u>display boards</u>.

<u>Prong paper fasteners</u> join pieces of <u>paper</u> and <u>card</u> together as <u>movable joints</u>.

<u>Hooks</u> can be used to <u>hang</u> materials — they're useful when creating displays.

<u>Staples</u> are a <u>permanent</u> or <u>temporary fixing</u> for paper or thin card. You can remove them with a staple-remover.

Bindings use Standard Components

Comb binding

1) <u>Holes</u> are punched in sheets using a special machine, then a <u>plastic comb</u> is inserted.

2) Pages can be added or removed <u>without causing damage</u>.

3) The book opens flat — <u>easy to read</u>.

Spiral binding

1) A <u>plastic coil</u> is inserted down the spine.

2) <u>Wiro binding</u> is similar but uses a <u>double loop wire</u> instead of a plastic coil.

Saddle stitching

1) <u>Double-sized pages</u> are folded and <u>stapled together</u> at the centre.

2) Easy and cheap — but it <u>won't hold many sheets</u>.

3) The books open more or less <u>flat</u>.

Comb binding

Spiral binding

Wiro binding

Standard Components

Some Bindings are **More Expensive**

Perfect binding

1) Pages are folded together in <u>sections</u>.
2) Each section is <u>roughened</u> at the fold and then <u>glued</u> to the <u>spine</u>.
3) You can bind <u>lots of sheets</u> but you can't open the book <u>flat</u>.

Thread sewing

1) This method is more <u>expensive</u>.
2) The pages are <u>sewn</u> together in sections, then a <u>soft cover</u> is <u>glued</u> on.
3) The pages are less likely to come <u>loose</u> than with perfect binding.

Case-bound / hard-bound

These books are like <u>thread-sewn</u> ones but with a <u>hard cover</u>.

Seals, **Tape** and **Adhesives** Keep Things **Attached** to Each Other

Envelope seals

1) <u>Self-seal</u> envelopes use two strips of <u>adhesive</u> that seal together on contact.
2) <u>Peel and seal</u> envelopes also have two strips of adhesive but one is <u>covered by a strip</u> (often made of paper). To <u>seal</u> the envelope, just remove the <u>covering strip</u> and <u>press</u> the two strips of adhesive together.
3) These seals are <u>very strong</u> and difficult to break. They are used for important or <u>confidential letters</u> because they can't be opened without tearing the envelope — so you can see if your letter has been <u>tampered with</u>.

Double-sided tape

1) <u>Double-sided tape</u> is useful for attaching two separate pieces of paper or board together <u>without a visible seal</u>.
2) This tape usually has a <u>peelable</u> layer on one side — this is removed when the tape is used.

Tabs

1) <u>Tabs</u> are stickers that can be used to seal something.
2) For example, tabs can be used to keep a leaflet or booklet <u>closed</u> while it <u>travels through the post</u>.

Adhesives

1) Adhesives (<u>glues</u>) can be used to <u>bond</u> paper and card — they can be <u>applied</u> in different ways.
2) <u>Glue sticks</u> uses a <u>solid</u> glue that can be <u>pressed</u> onto a material at a join. It's <u>non-toxic</u>, <u>cheap</u> and <u>environmentally-friendly</u>. Glue can also be applied as a <u>liquid</u>, e.g. using <u>squeezy glue pens</u>. In both cases, the glue is <u>clear</u> when <u>dry</u>.
3) You can also use <u>spray adhesive</u> for <u>mounting photos</u> onto paper or card — they cover large areas well and allow for <u>repositioning</u> (unless it says 'permanent').

You can find standard components all over the place...

You probably know about some of these standard components already, but bindings are a bit more unusual. It's not difficult to muddle up the different types, so make sure you know which is which.

Working with Paper and Board

OK, you've probably known about <u>scissors</u> since you were old enough to run with them, but that's no excuse to skip these pages. There are plenty of other, more <u>exotic</u> cutting tools out there.

Scissors can Cut Paper and Thin Card

1) You can cut <u>paper</u> and <u>thin card</u> well with scissors — but they're not much good for <u>fine detail</u> or for <u>removing</u> bits from within a sheet of paper or card (scalpels and knives are good for this — see below).

2) You can use '<u>pinking shears</u>' to produce an interesting <u>zigzag</u> edge (which is commonly used with fabrics — see p.99).

Knives and Cutters also Cut Card and Paper

1) There are loads of different <u>craft</u>/<u>trimming</u>/<u>hobby knives</u>.

2) You mainly use these to cut <u>card</u> and <u>paper</u> — though some will cut thicker board, balsa wood, etc.

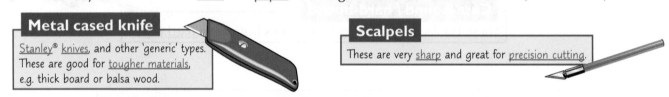

Metal cased knife

<u>Stanley®</u> <u>knives</u>, and other 'generic' types. These are good for <u>tougher materials</u>, e.g. thick board or balsa wood.

Scalpels

These are very <u>sharp</u> and great for <u>precision cutting</u>.

Plastic trimming knife

Similar to the metal cased knife, but some have <u>retractable blades</u> or <u>blade covers</u> for <u>safety</u> when not in use.

Circle cutters

1) You use these to cut <u>arcs</u> and <u>circles</u> in card and paper.

2) You can vary the <u>diameter</u> of the arc or circle to be cut.

Perforation cutters

1) These have a round blade which <u>rotates</u> as you push it along, making a line of lots of small <u>cuts</u>.

2) These can be used to make <u>tear-strips</u>, like the ones you find on forms where you're asked to detach a bit and send it back.

These are a bit like pizza cutters.

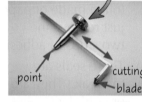

point   cutting blade

Circle cutters are used like a mathematical compass.

Guillotines and Paper Trimmers Cut Large Sheets

1) <u>Guillotines</u> are used to cut <u>large sheets</u> of paper and card, often <u>many sheets</u> at a time.

2) They have a <u>large blade</u> that you <u>push down</u> to produce a nice <u>straight cut</u>.

3) <u>Paper trimmers</u> are a similar piece of equipment that uses a <u>smaller rotary blade</u> to make straight cuts.

Equipment like guillotines, paper trimmers and die cutters (see next page) use shear forces (see p.45) to cut through paper and board.

Laser Cutters are Expensive but can Make Detailed Cuts

1) <u>Laser cutters</u> are machines that cut out designs drawn using <u>CAD</u>. They can make really <u>accurate</u> and <u>fine cuts</u> through paper and card.

2) They use a very fine laser beam to <u>burn away</u> material. This can be used to make intricately <u>detailed</u> decorative work that can be found on some <u>greetings cards</u> and <u>lampshades</u>. (There's more on laser cutters on p.5.)

Working with Paper and Board

Scoring Makes Paper and Board Easier to Fold

1) If you want to fold something <u>neatly and accurately</u>, then scoring the paper or board first will help.

2) <u>Scoring</u> makes a <u>small indent</u> on the page where it should be folded. To score by hand:

1) Draw a <u>guideline</u> with pencil where you want to fold, and line up a <u>metal ruler</u> along it.

2) <u>Run a knife</u> (e.g. a trimming knife) along the line, using the <u>ruler for guidance</u>.

3) Be careful to only <u>press lightly</u> with the knife to avoid cutting through the paper/board.

3) Scoring makes sure that <u>folds are straight</u> and <u>in the right place</u>.

4) Scoring can also be done using <u>CAD/CAM</u>, with only certain lines in a design being scored.

Die Cutting is Used for Cutting and Creasing Nets

1) Die cutting is used to <u>cut shapes out of</u> and <u>crease lines into</u> paper and board.

2) It's used to produce the nets of complex shaped products like <u>packaging</u>.

3) A <u>net</u> is a <u>2D</u> plan for making a <u>3D</u> object.

4) You can use CAD to design a net, and then use a <u>die cutter machine</u> to cut it out. These can be manual machines, operated by hand, or they can be computerised CAM machines.

5) The die itself (a bit like a giant cookie cutter) has <u>sharp blades</u> (for cutting) and <u>round-edged blades</u> (for creasing). These are formed into the right shape, so that they match the design of the net.

6) The die is mounted on a <u>strong plywood</u> base and <u>pressed down</u> onto the card.

7) You can cut through <u>many layers</u> of material at a time, so you can make <u>lots</u> of nets <u>very quickly</u>.

Cube net

1) A die cutter <u>presses</u> out the <u>net</u> from the sheet of material, using a <u>sharp blade</u> specially shaped to the outline of the net.

2) <u>Creases</u> can be made (along the lines where the material will be folded) by <u>rounded creasing bars</u>.

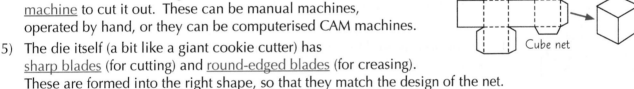

cutter   card   creasing bar

3) You have to make a <u>blade</u> especially to <u>match your net</u>, so die cutting is <u>expensive</u>, but it's great for making <u>large quantities</u> of nets with complicated designs.

8) Die cutting machines can be used for <u>loads of different things</u>, not just making nets. They can make <u>stencils</u> or cut out <u>particular shapes</u> and <u>designs</u>.

It's important to pick the right cutting tool for the job...

It's easy to forget the detail of all the different cutting methods covered on these two pages, especially when you're under pressure in the exam. To help get it clear in your mind, make a list of six different cutting tools and write down a suitable use for each tool.

Printing Techniques

I can tell you're desperate to learn all about the different <u>printing methods</u>. So here are two pages absolutely <u>jammed full</u> of fun printing facts. Who knew revision could be so <u>exciting</u>...

Lithography and Offset Lithography Use 'Oily' Ink

1) Lithography uses an <u>oil-based ink</u> and <u>water</u> and works on the principle that <u>oil and water don't mix</u>.

2) <u>Ultraviolet light</u> is used to transfer the image onto a smooth <u>aluminium printing plate</u> — the <u>image area</u> gets coated with a chemical that <u>attracts</u> the <u>oily ink</u> but <u>repels water</u>.

3) So the <u>image</u> area holds <u>ink</u> and the <u>non-image</u> area holds <u>water</u>.

4) In <u>offset lithography</u>, the image is printed onto a rubber '<u>blanket</u>' cylinder which squeezes away the water and transfers the ink to the paper.

5) Lithography and offset lithography are <u>fast</u> ways of printing and they give you a <u>high-quality</u> product.

6) They're great for print runs of <u>1000 copies or more</u> — so you can print <u>books</u>, <u>newspapers</u>, <u>magazines</u>, <u>packaging</u>, etc.

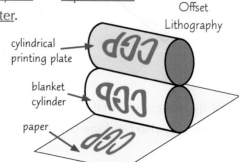

Offset Lithography

cylindrical printing plate

blanket cylinder

paper

Flexography Uses a Flexible Printing Plate

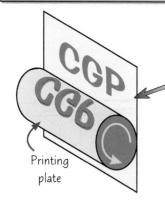

Printing plate

1) <u>Flexography</u> uses a printing plate made of <u>flexible rubber or plastic</u>. The image <u>sticks out a bit</u> from the plate.

2) You can print onto <u>different surfaces</u> using flexography — they don't have to be completely flat either. This means you can print onto things that aren't totally smooth, like <u>cardboard</u>, or other packaging such as <u>plastic bottles</u>.

3) It's <u>quicker</u> than lithography and the printing plates <u>last longer</u>.

4) Flexography is used for <u>large</u> print runs (over 5000) like <u>packaging</u>, <u>wallpaper</u> and also <u>carrier bags</u>.

Gravure Uses an Etched Printing Plate

1) Gravure uses an <u>etched</u> brass printing plate — meaning the image is <u>lower</u> than the surface of the plate and the ink fills the etched bits.

2) It's <u>expensive</u> to set up but it's <u>really fast</u> and ideal for <u>very large print runs</u> (a million copies or more). The products are <u>higher quality</u> than ones printed using lithography.

3) Gravure is used to make products such as <u>postage stamps</u>, <u>photos</u> in books and catalogues, and <u>colourful magazines</u>.

Screen Printing Uses... a Screen

1) In screen printing, a <u>stencil</u> is put under a <u>fine mesh screen</u>, and <u>ink</u> is spread over the top. The ink goes through the stencil and prints onto the material below.

2) It's a <u>low-cost</u> process, ideal for <u>short</u> print runs of up to a few hundred copies where <u>fine detail</u> isn't needed.

3) You can use it to print onto <u>various surfaces</u> (e.g. paper, card, fabric) — so it's great for printing <u>posters</u>, <u>estate agents' signs</u>, etc. It can also be used on <u>textiles</u> — take a look at p.106.

Printing Techniques

Digital Printing Doesn't Use Printing Plates

1) Digital printing is done using <u>inkjet</u> and <u>laser</u> printers.

2) You don't have to make any printing plates, so it's <u>less fiddly</u> than many other methods.

3) There are <u>no set-up costs</u> apart from buying a printer and ink/ toner cartridges (which will need replacing when they run out).

4) Digital printing is <u>expensive</u> per sheet but for <u>short</u> print runs (hundreds of copies) it's <u>cheaper</u> than setting up the plates for another printing process, e.g. lithography.

5) It's used to print <u>posters</u>, <u>flyers</u>, <u>digital photos</u>, etc.

Digital Printers Use Four Colours in Layers

1) Digital printers use four colours — <u>cyan (C)</u>, <u>magenta (M)</u>, <u>yellow (Y)</u> and <u>black (K)</u>. These are also referred to as CMYK and are known as <u>process colours</u>.

2) Anything that's printed in colour is made up of a <u>mixture</u> of these colours.

> The 'K' stands for <u>'key'</u> — it means black. You can make black by <u>mixing</u> the other three colours, but using <u>black ink</u> usually looks better, and it works out <u>cheaper</u> if you're printing a lot of black.

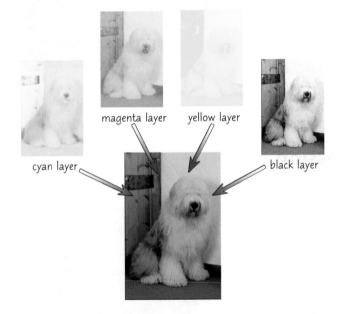

magenta layer   yellow layer

cyan layer   black layer

3) When the computer is instructed to print, the printer recognises the required colour and adds <u>layers</u> of cyan, magenta, yellow and black to make the final colour.

4) Some printers use special <u>spot colours</u> (e.g. PANTONE® colours) as well — to print particular colours that <u>can't be achieved with CMYK</u>.

5) There's no set <u>order</u> in which the colours are put onto the paper. Most printers stick to the CMYK order while others go from <u>lightest to darkest</u> — YMCK.

Lithography Also Uses These Four Colours

1) Lithography (see previous page) uses <u>four</u> aluminium <u>printing plates</u> — that's one plate for each of the <u>CMYK</u> colours.

2) The image is built up in <u>layers of colour</u> (as above).

Remember — there's upsides and downsides to each technique...

There's quite a lot to learn on these couple of pages. Now you know everything you need to know about printing techniques, and why pictures look funny when you're running low on ink — it's all down to CMYK.

Paper and Board Finishes

Once you've printed your product, you can choose a <u>finish</u> to... you've guessed it... finish your product. There are lots of different finishes to choose from, so you need to think about which <u>properties</u> you'll need.

Print Finishes can Make Your Product Look Top Quality

After the colours have been printed, you can use a <u>finish</u> on your product.

1) Finishes can <u>improve</u> the <u>look</u> of your image. Your final product will look nice and professional — so people will be more likely to <u>buy it</u>.
2) They help to <u>protect</u> your product from being damaged.
3) But adding a print finish can be <u>expensive</u>.

Laminating Means Sandwiching in Plastic

1) Laminating means <u>sandwiching</u> a document, e.g. a menu, <u>between</u> two layers of <u>plastic</u>.
2) The laminating machine <u>heats</u> the plastic and <u>seals</u> it together.
3) Laminating <u>business cards</u>, <u>menus</u> and <u>posters</u> makes them last longer without getting <u>damaged</u>.
4) It's a <u>quick and effective</u> way to finish a piece of work on paper or thin card.
5) Many <u>packaging materials</u> are a lamination of different papers, cards, plastics and aluminium foil.
6) Laminating can also apply plastic to <u>just the printed surface</u> — i.e. only one side of the product.
7) It can give a <u>shiny</u> or a <u>matt</u> finish.

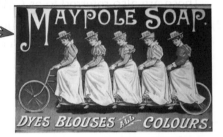

Embossing Leaves a Raised Impression

1) <u>Embossing</u> means pushing a <u>shaped die</u> into the back of the material to leave a slightly <u>raised impression</u> on its surface.
2) It's often used to <u>draw attention</u> to a particular bit of a product, e.g. the <u>title</u> of a book, a <u>logo</u> or an <u>image</u>.
3) It's an <u>expensive</u> process but it adds <u>texture</u> and can suggest <u>quality</u>.
4) Industrial embossing machines use a <u>rolling die</u> to emboss the <u>same pattern</u> onto <u>large amounts</u> of paper and card.

Foil Application Makes Things Look Fancy

1) Foil application (or foil blocking) means using <u>heat</u> and <u>pressure</u> to <u>print metal foil</u> onto certain areas of a product.
2) Like embossing, it's used in packaging to <u>draw attention</u> to a <u>logo</u> or <u>brand name</u>, and to give the impression of a <u>quality</u> product — but it's <u>expensive</u>.
3) It's also used on <u>greetings cards</u>, <u>book titles</u> and <u>wrapping paper</u>.

Print finishes aren't always used — they can be quite expensive...

So now you know how products get those fancy finishes, but you're not done yet — there's more to learn...

Paper and Board Finishes

Varnishing Makes Things **Shiny** or **Matt**

1) Varnishing is used to make things look <u>smooth</u> and <u>glossy</u> or <u>matt</u>, so they look more exciting and high-quality.

2) You can varnish the <u>whole product</u>, e.g. <u>playing cards</u> — this makes them slide over each other.

3) Or you can varnish <u>specific areas</u> (e.g. <u>titles on book covers</u>) to draw attention to them — this is called <u>spot varnishing</u>.

4) Varnish <u>can't be written on</u>, so varnish is sometimes <u>only applied to one side of the paper</u> — e.g. postcards.

UV Varnishing

1) In <u>UV varnishing</u>, a varnish is <u>applied to the surface</u> of the paper. The varnish is then <u>cured under UV light</u> — this makes it feel <u>dry to the touch</u>.

2) A <u>glossy</u>, shiny finish or a <u>matt</u> finish can be applied.

3) UV varnishing can be applied to the <u>whole page</u> (<u>flooding</u>) or to <u>certain areas</u> (<u>spot UV varnishing</u>).

4) UV varnishing is a good finish for items that are going to be <u>handled frequently</u> as it protects the paper. It's often used for <u>business cards</u> and <u>magazine covers</u>.

5) The <u>varnish cures very quickly</u> when it is exposed to UV light. This <u>speeds up</u> the printing process and <u>reduces production times</u>.

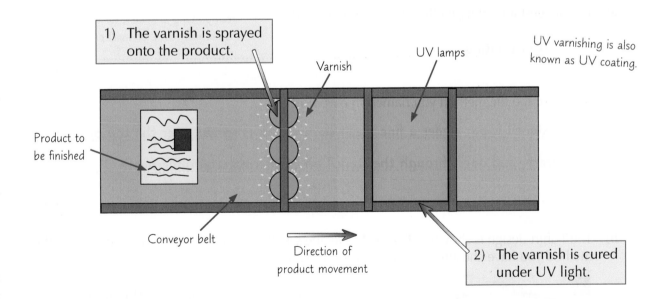

1) The varnish is sprayed onto the product.

Varnish

UV lamps

UV varnishing is also known as UV coating.

Product to be finished

Conveyor belt

Direction of product movement

2) The varnish is cured under UV light.

Varnishes can be used to make a product look fancy...

EXAM TIP

There are a few ways of finishing a product, so make sure you learn them all. You could be given a product in the exam and asked to name or describe a finish that would be suitable for it. When answering a question like this, you need to think about what the product is and how it'll be used — e.g. if it's going to be handled lots then the protection offered by lamination might be ideal.

Warm-Up and Worked Exam Practice

Now that you've learned all there is to learn about paper and board, it's time to check that it's all sunk in.
Thankfully, there is a whole batch of questions coming up that I've hand-picked just for you...

Warm-Up Questions

1) Paper can be sold in rolls. True or false?
2) Give two examples of standard components that can be used with paper or board.
3) Name a machine that could be used to mass produce a net.
4) Name the process of sandwiching a document between two layers of plastic.

Worked Exam Questions

1   Paper products can have intricate designs cut into them.
An example is the greetings card shown in **Figure 1**.

These designs can be drawn using computer aided design (CAD) software.
Name a machine that is capable of cutting out such a design,
and describe how it is done.

A laser cutter is capable of cutting this design. It uses a very

fine laser beam to burn away parts of the paper.

[2 marks]

Figure 1

2   A design is going to be printed onto a small batch of T-shirts.

    a) **i)** Suggest a printing method that could be used.

          Screen printing is suitable for this job as
          it is normally used for small print runs.

        Screen printing

[1 mark]

      ii) Describe the method you named in part **a) i)**.

        A stencil is put under a fine mesh screen. Ink is spread over the top of the

        screen, and goes through the stencil on to the material underneath.

[2 marks]

    b) The T-shirt design is shown in **Figure 2**. Decide whether or not this design is suitable for the
method you have chosen and explain your answer.

Figure 2

    Screen printing is not suitable for this design.

    The design is too complicated to be printed using

    a screen and stencil.

[2 marks]

Exam Questions

1 Standard components are pre-manufactured parts.

Describe the benefits to manufacturers of using standard components in the manufacturing process.

....

....

....

[2 marks]

2 A couple are sending out 100 wedding invitations to their guests. The invitations are to be printed in full colour with foil applied to the names of the couple.

The table below shows two quotes for printing the invitations from a local printing company.

a) Complete the table by calculating the price per invitation for each of the quotes.

	Features of the invitation	Price for 100 invitations	Price per invitation
Quote 1	Full colour + foil application	£180	
Quote 2	Full colour + foil application + embossing	£340	

[1 mark]

b) Calculate the extra cost per invitation of embossing.

....

[1 mark]

c) Calculate the percentage increase between quote 1 and quote 2.
Give your answer to 1 decimal place.

....

....

[2 marks]

3 "The best cardboard for packaging products (e.g. boxes) is strong and highly rigid."
Evaluate this statement.

In your answer, you should talk about how you agree and disagree with the statement.

....

....

....

....

....

....

....

[4 marks]

Revision Questions for Section Four

Congrats, you've reached the end of <u>Section Four</u>. Well, nearly — just a few lovely questions to go...
* Try these questions and <u>tick off each one</u> when you <u>get it right</u>.
* When you've done <u>all the questions</u> for a topic and are <u>completely happy</u> with it, tick off the topic.

Properties of Paper and Board (p.62-63) ☑

1) Why might recycled paper or board not be suitable for use in food packaging?
2) Give two properties that are useful for food packaging to have.
3) Why does the print quality not need to be high for flyers and leaflets?
4) Why are additives added to paper and board?
5) How much larger than A4 is A2?

Standard Components (p.64-65) ☑

6) What are standard components?
7) Which of the following components is often used to attach paper or card to a display board?
   a) treasury tags   b) drawing pins   c) plastic comb   d) tabs
8) Lauren is choosing a binding for a book. It is a small book without many pages. It needs to be cheap to produce, and must lie flat when opened. Name a suitable type of binding.

Working with Paper and Board (p.66-67) ☑

9) Describe what the following equipment could be used for:
   a) Circle cutter
   b) Perforation cutter
   c) Scalpel
   d) Guillotine
10) What is scoring used for?
11) Give one disadvantage of using die cutting.

Printing Techniques (p.68-69) ☑

12) Which printing method uses a plate made of rubber or plastic?
13) Name the two printing techniques that are suitable for printing magazines.
14) Why is digital printing used for short print runs?
15) What colours do the letters CMYK stand for?

Paper and Board Finishes (p.70-71) ☑

16) Name a product that might use the following finishes and suggest why.
   a) laminating
   b) foil blocking
17) a) What is embossing?
    b) Why might it be used?
18) Why is varnish sometimes only applied to one side of paper?
19) a) Name the varnishing process that uses UV light.
    b) Suggest why using UV light is useful when producing varnished products on a large scale.

Uses of Wood, Metals and Polymers

So here it is, the section that's full of info about <u>wood</u>, <u>metals</u> and <u>polymers</u>...

Look at the **Properties** of a **Material** to See What it'll be **Good** at

1) As you know from <u>Section Two</u>, different <u>materials</u> have different physical <u>properties</u>.

2) When <u>making</u> a product for a <u>specific purpose</u>, the <u>materials</u> that it's made from need to have <u>properties</u> that are <u>suitable for that purpose</u>.

3) Here are some examples of <u>products</u> made from <u>wood</u>, <u>metal</u> and <u>polymers</u> and some of the reasons why these materials are <u>suitable choices</u>...

Have a look back at pages 18-19 for a reminder of some of the different properties materials can have.

Traditional wooden toys

1) <u>Wooden toys</u> are often made from <u>hardwoods</u> such as <u>beech</u>, <u>oak</u> and <u>maple</u>.

2) These woods are <u>hard</u> and <u>durable</u>, so they can <u>withstand a lot of force</u>. This makes them <u>safe for children</u> as they won't break and splinter.

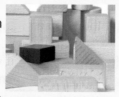

3) These woods have an <u>attractive appearance</u> or they can be painted in <u>bright colours</u>.

Flat pack furniture

1) <u>Flat pack furniture</u> is often made from <u>composites</u> such as <u>MDF</u> (see p.27).

2) MDF is <u>strong</u> so it can <u>withstand</u> the <u>forces</u> that are applied to furniture.

3) MDF has a <u>uniform texture</u> (no knots or grain) and it can be <u>shaped</u>. This is ideal for flat pack furniture which needs to be <u>cut to shape</u>, drilled in <u>specific locations</u>, etc.

4) MDF can also have <u>finishes applied</u> to it, resulting in <u>attractive</u> furniture.

Cooking utensils

1) <u>Cooking utensils</u> can be made from <u>stainless steel</u> (see p.22). Stainless steel is <u>strong</u> and <u>doesn't easily rust</u>. It also has a <u>high melting point</u> and is <u>safe for use with food</u>.

2) This makes it suitable to use in the <u>kitchen</u>, where it will be exposed to <u>moisture</u>, <u>heat</u> and <u>food</u>.

There are lots of other reasons why a material may be chosen for a product (see p.43-44). E.g. one of the main advantages of MDF is that it's cheap.

Hand tools

1) <u>Hand tools</u> (e.g. saws, chisels, etc.) are typically made from <u>tool steel</u> (a steel containing up to around <u>1.5% carbon</u>).

2) Tool steel is <u>hard</u>, <u>tough</u> and <u>strong</u>.

3) These properties make <u>tool steel</u> suitable for hand tools as they can <u>cut other materials</u> without breaking.

Polymer seating

1) Seats like these are often made from <u>thermoforming plastics</u> like <u>polypropylene</u>.

2) Polypropylene is available in lots of <u>colours</u> and is <u>easily moulded</u> into <u>comfortable</u> seat shapes.

3) It's quite <u>tough</u>, <u>flexible</u> and <u>strong</u>, so polypropylene seats can <u>withstand people sitting down</u> on them over and over again.

4) Polypropylene is <u>resistant to moisture</u> so it's suitable for <u>outdoor use</u> (however it can be affected by <u>UV light</u> — see the next page).

Electrical fittings

1) As you saw on p.23, <u>electrical fittings</u> are often made from <u>urea formaldehyde</u>.

2) This <u>thermosetting plastic</u> can be <u>moulded</u> into the <u>required shape</u>.

3) It'll stay <u>hard</u> and <u>rigid</u> even if it's heated again. This means it's <u>heat</u> and <u>fire-resistant</u>.

4) It's also an <u>electrical insulator</u>, which is essential for <u>safety reasons</u>.

Different materials can have very different properties...

Sometimes it can be tricky to decide which material is fit for a purpose, but you can often just use common sense — for example, you wouldn't make a light switch out of an electrical conductor...

Uses of Wood, Metals and Polymers

The **Properties** of Materials can be **Modified**

1) Sometimes materials <u>don't</u> have the <u>exact properties</u> that you want them to. Annoying.
2) Thankfully the properties of some materials can be <u>modified</u>.
   This can make them suitable for <u>different purposes</u>.
3) You can modify the properties of some <u>woods</u>, <u>metals</u> and <u>polymers</u>. Here are a few examples:

Wood can be **Seasoned** to **Strengthen** it

1) As you saw on page 55, <u>seasoning wood</u> involves <u>drying</u> it.
   It takes place after trees have been <u>cut down</u> and had the <u>bark removed</u>.
2) <u>Reducing</u> the amount of <u>moisture</u> in the wood makes it <u>stronger</u>.
   It also makes it less likely to <u>rot</u> or <u>twist</u>.
3) If wood is <u>not seasoned</u> before it is used in a product then it can <u>change shape</u>.

> For example, if you put unseasoned wood in a <u>dry environment</u>, it will <u>dry too quickly</u>.
> This can cause the wood to <u>shrink</u>, which can <u>damage the wood</u> and <u>reduce its strength</u>.

4) <u>Unseasoned wood</u> is often used for <u>timber-framed buildings</u> though. This is because it is <u>easier to work</u>, it's <u>cheaper</u> and can be sourced in a <u>wider variety</u> of <u>sizes</u> than seasoned wood.

Annealing can Make **Metals** More **Malleable**

1) <u>Heat-treating</u> metals can <u>change</u> their properties.
2) For example, <u>annealing</u> involves <u>heating</u> a metal (often until it <u>glows red</u>) and leaving it to <u>cool slowly</u>.
3) This makes the metal <u>softer</u>, <u>more ductile</u> and <u>less brittle</u>.
4) Softening the metal makes it more <u>malleable</u>.
   This makes it easier to <u>bend</u> and <u>shape</u>.

Stabilisers can **Protect Polymers** from **UV Light**

1) <u>UV light</u> from the sun or artificial light sources can <u>damage polymers</u> by causing <u>changes</u> to their <u>chemical structure</u>.
2) Polymers exposed to UV may <u>change colour</u>, <u>fade</u>, <u>lose their strength</u> and become <u>brittle</u>.
   They can also get a <u>powdery residue</u> on their surface — this is known as '<u>chalking</u>'.
3) This can be <u>prevented</u> by adding a <u>chemical</u> called a <u>UV stabiliser</u> to the polymer — this <u>protects</u> the <u>structure</u> from <u>UV</u>.
4) This means the polymer can be used <u>outside</u>, e.g. for garden furniture and stadium seating.

Seasoning means removing some of the moisture from wood...

Sadly, it's not enough to know all the fancy words for the exam, you need to know what they mean too. A good way to do this is to write down the terms you need to know (e.g. annealing), try and write down what they mean and then check your answers against the book. Repeat until it sticks.

Stock Forms and Standard Components

Timber, <u>metal</u> and <u>polymers</u> can be <u>bought</u> in different <u>shapes</u> (known as <u>stock forms</u>) and <u>sizes</u>.

Timber is Available in **Planks**, **Strips**, **Mouldings** and **Boards**

Timber can be bought in lots of forms:

1) Timber comes in different <u>sizes</u> — for example <u>planks</u> and <u>strips</u> are sold in <u>standard thicknesses</u>, <u>widths</u> and <u>lengths</u>, e.g. 47 x 100 x 2400 mm.

2) <u>Planed Square Edge</u> (<u>PSE</u>) timber has its rough surfaces shaved off with an electric <u>planer</u> to give <u>smooth surfaces</u> and <u>sharp corners</u>.

3) <u>Rough sawn</u> timber is <u>not smoothed</u> after it's cut — so it's <u>cheaper</u> and useful for construction work where it <u>won't be visible</u>.

4) <u>Mouldings</u> are strips that come in a range of <u>cross-sections</u>. They're used for skirting boards, door frames, picture frames, etc.

5) <u>Manufactured boards</u> (MDF, plywood, blockboard, etc.) are usually available in <u>2440 x 1220 mm sheets</u> (although you can get smaller sizes). <u>Common thicknesses</u> are 4, 6, 9, 12, 15 and 18 mm.

Remember, timber is wood that has been turned into a useful form.

Mouldings with circular cross-sections are available in different diameters.

diameter

Metals Come in Loads of **Shapes** and **Sizes** too...

1) Metals are commonly available in a <u>wide range</u> of shapes and sizes, because it can be very difficult to convert one shape to another.

2) This means that the <u>manufacturers</u> can buy roughly the right shape to start working with.

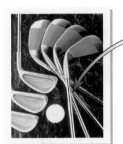

For example, the shafts of these <u>golf clubs</u> were made by using <u>tubes</u> of metal.

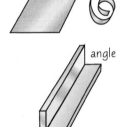  

sheet
strip
bars
rod
tube
angle
U-shaped channel
I-shaped girder

Like timber, <u>metals</u> are available in a range of <u>thicknesses</u>, <u>widths</u> and <u>lengths</u>. Lengths of metal with a <u>circular cross-section</u> can be bought in different <u>diameters</u>.

...and so do **Polymers**

Here are some of the <u>different forms</u> that <u>polymers</u> come in...

You buy granulated and powdered polymers by weight.

 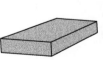  

<u>Sheets</u>, <u>tubes</u> and <u>rods</u> can be cut to size and bent (see p.88). Sheets can also be used in drape forming (see p.90).

<u>Foam</u> is used for <u>protective packaging</u> and making <u>models</u> and <u>mock-ups</u>.

<u>Films</u> are good for <u>vacuum forming</u> (p.89), and for use as <u>windows</u> in packaging.

<u>Granules</u> can be <u>melted down</u> and used in <u>casting</u> and <u>moulding</u> (see p.88-89).

<u>Powders</u> can be used in some types of <u>moulding</u>, as <u>coatings</u> (see p.92) or in <u>3D printing</u> (see p.87).

Some polymers (e.g. foam and rods) are available in standard <u>widths</u>, <u>lengths</u> and <u>thicknesses</u> (or <u>lengths</u> and <u>diameters</u>). Some <u>thin polymers</u> (e.g. sheets and films) are available in different <u>gauges</u> (<u>standard thicknesses</u>) — gauge gives an indication of how <u>strong</u> the polymer will be.

Stock Forms and Standard Components

You need to know about some <u>pre-manufactured standard components</u> (see p.64) and how they can be used.

Screws and Bolts are Temporary Fastenings

1) <u>Screws</u> and <u>nuts and bolts</u> are <u>temporary</u> ways of joining things together — you can take them <u>apart</u> if you need to.

2) They are usually made from steel, brass or stainless steel, and are '<u>self-finished</u>', plated with <u>zinc</u>, <u>brass</u> or <u>chrome</u>, or coated in a <u>black varnish</u>.

3) It's important to use the <u>right sort</u> to make sure that the job <u>looks good</u> and <u>holds together properly</u>.

Temporary fixings are designed to be taken apart, and put back together if necessary. Permanent fixings are... permanent.

Screws

1) Screw threads (the twisty bit around the outside of the screw) <u>grip</u> the material, making a <u>strong</u>, <u>tight fixing</u> in wood, metal and plastic.

2) They are used for lots of jobs, e.g. for access to the batteries in <u>children's toys</u>, <u>self-assembly furniture</u>, fixing <u>shelves</u> and <u>mirrors</u> to walls, and fixing <u>hinges</u> to doors.

There are <u>different types</u> of screws for use with <u>wood</u>, <u>metals</u> and <u>plastics</u>:

Screws can be used with glue if you want a permanent fixing.

1) <u>WOODSCREWS</u> often require '<u>pilot</u>' holes to be drilled before the screw is inserted.

2) As the screw is turned by a <u>screwdriver</u>, the thread pulls it into the wood.

3) Different types of <u>head</u> are available for different jobs, e.g. <u>round</u>, <u>countersunk</u>, <u>slotted</u> and <u>cross heads</u>.

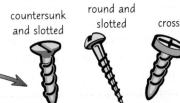

countersunk and slotted   round and slotted   cross

1) <u>SELF-DRILLING SCREWS</u> have <u>hardened threads</u> and a <u>tip</u> like a <u>drill bit</u>.

2) They are designed to <u>drill their own threaded holes</u> in hard materials such as <u>metals</u> and <u>hard plastics</u>. They don't require a <u>pilot hole</u> to be drilled before they're inserted.

1) <u>MACHINE SCREWS</u> have a straight (not tapered) <u>threaded shank</u> and are used with <u>washers</u> and <u>nuts</u> or in a hole with <u>matching threads</u>.

2) Their heads vary (round, countersunk, etc.). Some are tightened with a <u>screwdriver</u> (cross and slotted types) and some with an <u>Allen key</u> (socket head).

3) They are often used to join <u>metal</u> parts together.

Straight shank

Nuts and Bolts

1) Nuts and bolts are used to join <u>thin materials</u> like sheets of <u>metal</u> or <u>plastic</u>.

2) They're only useful if you can get to <u>both sides</u> of the materials you're joining.

3) They're used to join parts of <u>machinery</u>, e.g. in cars and bridges, and to hold <u>moving parts</u>, e.g. in play swings.

4) <u>Bolts</u> are similar to machine screws (see above) but usually have a <u>square</u> or <u>hexagonal</u> head that can be tightened with a <u>spanner</u>.

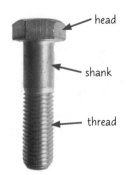

bolt

nut

head

shank

thread

Most materials are commercially available in stock forms...

Draw a mind map to show the stock forms that timber, metal and polymers can be bought in.

More Standard Components

More components for you now. First up, rivets and hinges, then onto knock-down fittings...

Rivets are Mainly Used for Joining Sheet Metal

1) A rivet is a metal peg with a head on one end.
   Rivets are mostly used for joining pieces of metal.

2) A hole is drilled through both pieces of metal and the rivet is inserted
   with a set (hammer-like tool). The head is held against the metal whilst
   the other end is flattened and shaped into another head with a hammer.

3) 'Pop' (or 'blind') rivets are now very common. They can be used where there is only access to
   one side of the material (hence 'blind' rivet). It's a fast and easy method of joining sheet metal.

standard rivets

Rivets are permanent fixings.

How pop rivets work

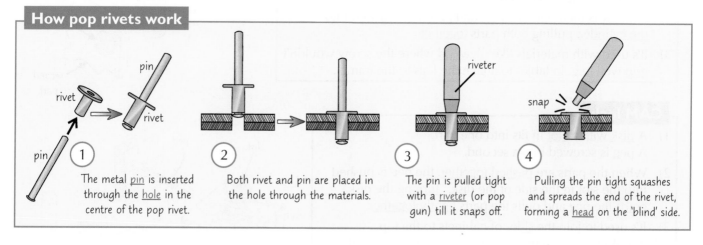

pin

rivet

rivet

rivet

pin

1) The metal pin is inserted through the hole in the centre of the pop rivet.

2) Both rivet and pin are placed in the hole through the materials.

riveter

3) The pin is pulled tight with a riveter (or pop gun) till it snaps off.

snap

4) Pulling the pin tight squashes and spreads the end of the rivet, forming a head on the 'blind' side.

There are Four Main Types of Hinge

Hinges are available in steel, brass and nylon (a polymer), and can be coated to match a piece of furniture. The part of the hinge that moves is called the knuckle.

Butt hinges

1) Butt hinges are the most common type of hinge used for doors.
2) One part of the hinge is set into the door and the other part is set into the frame.
3) They're available in brass or steel.

Tee hinges

1) Tee hinges are often used outside for things like shed doors or garden gates. The longer 'strap' allows the hinge to support a greater weight.
2) They're often covered in black enamel.

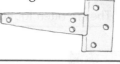

Pivot hinges

1) Pivot hinges allow you to lift a door off its frame.
2) The hinge is made from two parts that fit together. One part is screwed to the door and the other is screwed to the door frame.

Flush hinges

1) Flush hinges are screwed directly onto the surface of the wood, so they're easier to fit than butt hinges.
2) They're usually used for lightweight jobs.

The Eiffel Tower contains 2.5 million rivets...

Rivets are still a popular way of joining metals together today. Hinges are another really common standard component — they come in various shapes and sizes and each type is suited to a different job.

More Standard Components

Knock-Down (KD) Fittings are Temporary Joints

1) Knock-down fittings are <u>temporary</u> fittings that allow furniture to be <u>assembled</u> and <u>taken apart easily</u>. They're usually used for cheap 'flat-pack' furniture.

2) They're very <u>fast</u> to use, but are nowhere near as <u>strong</u> as glued joints.

3) Most types are assembled with <u>screwdrivers</u> or <u>Allen keys</u>.

4) There are different types for <u>different joints</u>...

Scan fittings

1) A <u>cylinder</u> with a <u>screw thread inside</u> is put into a ready-made <u>hole</u> in a part.

2) A <u>screw</u> is then put into <u>a second part</u> and screwed <u>into the cylinder</u>, pulling both parts <u>together</u>.

3) It's used with materials like <u>plywood</u> where the screw wouldn't <u>grip</u> well (e.g. in tables to attach the legs to the frame).

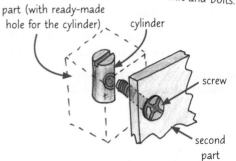

Scan fittings are sometimes called barrel nut and bolts.

part (with ready-made hole for the cylinder) — cylinder — screw — second part

CAM locks

1) A <u>disk</u> with a <u>slot</u> in fits into one part. A <u>peg</u> is screwed into a second.

2) When the parts are <u>pushed together</u>, the peg is pushed into the slot in the side of the disk. <u>Turning</u> the disk <u>grabs</u> the <u>peg</u> and pulls the parts <u>tightly together</u>.

3) It's used to join the <u>sides</u> of cabinets to the <u>top</u>. It looks quite <u>neat</u> (it's mostly hidden <u>inside</u> the parts).

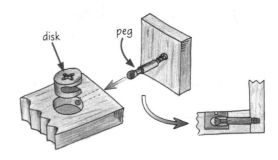

disk   peg

Single and two-piece blocks

1) These are <u>plastic blocks</u> used to join <u>parts</u> at <u>right angles</u>.

2) They have holes in them to fit <u>screws</u> through, and screw into ready-made <u>holes</u> in the parts being joined.

3) <u>Two-piece blocks</u> are first screwed into the parts, and then joined together with a <u>bolt</u>, making the joint straight and strong.

4) They can be easier to attach in <u>tight corners</u> than single piece blocks, and can be <u>undone</u> if required.

5) They're often found holding up <u>shelves</u> in wardrobes.

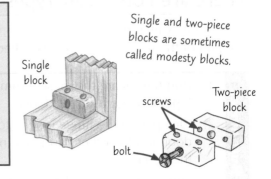

Single and two-piece blocks are sometimes called modesty blocks.

Single block   screws   Two-piece block   bolt

Dowels

1) Dowels are <u>wooden rods</u> that are pre-cut to length.

2) They're used in <u>dowel joints</u>. Holes are drilled into the <u>ends</u> and <u>sides</u> of <u>parts</u>, and the dowels fit into them like a <u>peg</u>.

3) They're usually used with <u>glue</u>, and give more <u>contact area</u> to the joint to make it stronger. The dowels have little <u>channels</u> in them to make space for the glue. Gluing makes them a <u>permanent fixing</u>.

4) They're often used to attach <u>shelves</u> to cabinets.

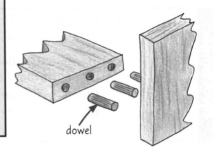

dowel

Knock-down fittings allow furniture to be produced cheaply...

Knock-down fittings aren't the most exciting thing you'll learn about, but you might need to write about them in detail in the exam — make sure you learn how they work and what they're used for.

Shaping Materials — Hand Tools

Wood, metals and polymers can be cut, shaped and smoothed. This can be done using hand tools. These are (surprisingly) tools that work by hand (i.e. not machine tools or power tools).

Saws are the Main Cutting Tools

1) There are different saws for different materials. Saws with lots of little teeth close together (fine pitch) are usually for hard materials and saws with bigger teeth further apart (coarse pitch) are for soft materials. Using the wrong saw could damage the material (or hurt you — it could 'jump' or catch unexpectedly).

Rip saw — for cutting wood along the grain

Tenon saw — for making straight cuts in small pieces of wood

Hacksaw — for metals and plastics

Coping saw — for cutting curves in wood or plastic

2) Tips for using a saw:

- Try to use the whole length of the blade so that some teeth don't wear out faster than others.
- Don't press too hard or the saw might jam in the material. This will damage the blade, the material or even you when you try to pull it out.
- Rough edges from sawing can be tidied up by sanding or planing.

Sounds obvious, but saws need to be kept sharp — either by sharpening or replacing the blade.

Chisels are Used for Shaping

Chisels are used to cut away and shape wood and metal.

1) Wood chisels come in different profiles for making different shapes. You hit them with a mallet.
2) Gougers are used for sculpting.
3) For metal, you need cold chisels. These are hit with a hammer.

Planes, Files and Abrasive Papers can Shape and Smooth

1) A bench plane has an angled blade that shaves off thin layers of material. It's used on wood for removing material (shaping).

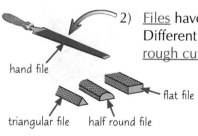
hand file
flat file
triangular file   half round file

2) Files have hundreds of small teeth to cut away at a material. Different 'cuts' make them suitable for different processes: rough cuts are for removal of material, fine cuts are for finishing (final smoothing).

They come in different profiles — to make different shapes. Most files are meant for metals and wood. There are special ones with very coarse teeth called cabinet rasps that can be used on wood.

3) Abrasive papers such as sandpaper (or glass paper) are rubbed against the material you want to shape or smooth. They are often wrapped around a sanding block so an even pressure can be applied and to give you better grip. Abrasive papers are available in a range of grit sizes — from coarse to fine. Different types are available for use on wood, metal, and polymers.

Wet and dry (silicon carbide) paper is a type of abrasive paper that is often used on plastics and metals. It can be used wet — this stops it getting full of dust.

Shaping Materials — Hand Tools

Drills Make Holes

1) A <u>bradawl</u> can be used to help you drill in the right place. You <u>press</u> it into the material to make a <u>little dent</u> where you're going to drill — this <u>stops</u> the <u>drill bit</u> from <u>slipping</u>. Bradawls can only be used on <u>wood</u> and <u>plastic</u>.

2) A <u>centre punch</u> is used for <u>metal</u> — you <u>hit</u> it with a <u>hammer</u> to make a dent in the metal's surface.

3) Depending on how hard the material is, you can do the actual drilling with a <u>brace</u>, a <u>hand drill</u> or a <u>power drill</u>.

4) All drills work by rotating a <u>drill bit</u> against the material.

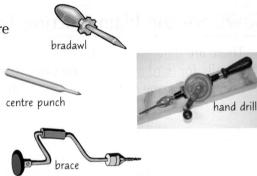

bradawl

centre punch

hand drill

brace

 <u>Twist bits</u> are used to drill <u>small holes</u> in wood, metal or plastic. <u>High speed steel</u> (HSS) twist bits are used on metals and plastics.

 <u>Flat bits</u> are used on <u>wood</u> and <u>plastics</u> to drill <u>large</u>, <u>flat-bottomed</u> holes. <u>Forstner bits</u> make similar holes, but can only really be used on <u>machine</u> drills (see p.86).

 <u>Countersink bits</u> make holes for <u>countersunk screw heads</u> to sit in.

 <u>Hole saws</u> are a bit like <u>round saws</u> that are used in a drill. They're used to make <u>big holes</u> in <u>thin material</u>.

5) Drill bits can get really <u>hot</u>, so be <u>careful</u> when taking them out of the material.

Hand Tools Need to be Used Safely

When you're using hand tools:

- Make sure the object you're working on <u>can't move</u> — hold it firmly in a <u>bench hook</u>, <u>vice</u> or <u>clamp</u>.
- Keep your <u>fingers</u> out of the way.
- Be aware of any <u>exposed blades</u>, and always carry tools <u>safely</u>.
- It's a good idea to wear <u>goggles</u> to stop <u>shavings</u> getting in your <u>eyes</u>.

Clamps are great for keeping an object steady whilst you're working on it.

REVISION TASK

So you can cut, shape and put holes in materials using hand tools...

It's important to select the right tool for the job and the material that you're using. If you don't, it's likely that you'll end up not using the tool properly to try and get it to do what you want — and that can be dangerous. Draw a table showing the different hand tools and what they're used for.

Warm-Up and Worked Exam Questions

That's the first half of Section Five done. There's a fair bit of info covered on those eight pages, so have another flick through and make sure you're happy with it. Once you've done that, give these questions a go.

Warm-Up Questions

1) Suggest a suitable thermosetting plastic that a plug socket could be made from.
2) Give one form of polymer that could be used for moulding.
3) What is a knock-down fitting?
4) Name a type of saw used for cutting curves in wood or plastic.

Worked Exam Questions

1 **Figure 1** shows the moisture content of two logs that have been cut from the same tree.

Log	Moisture content (%)
A	49
B	23

Figure 1

a) Which of the logs do you think has been seasoned? Give a reason for your answer.

Log B because it has a lower moisture content.

[1 mark]

b) Give **two** effects that seasoning has on wood.

Any other reasonable answer would be fine here e.g. it makes it less likely to rot.

1. It makes it stronger.

2. It makes it less likely to twist.

[2 marks]

2 Metals can be supplied in different shapes and sizes.

Figure 2

Figure 3

Suggest what shape of metal a manufacturer would use to make the following products:

a) the car bonnet shown in **Figure 2**.

sheet

[1 mark]

b) the bicycle frame shown in **Figure 3**.

rod

Tube metal would also be suitable.

[1 mark]

c) Explain why metals being available in a range of shapes and sizes is useful for manufacturers.

It can be very difficult to convert one metal shape to another. If there is a wide range of shapes and sizes available it means that manufacturers can buy roughly the right shape to start working with.

[2 marks]

Exam Questions

1   A carpenter is making an oak key rack,
    shown in **Figure 4**.

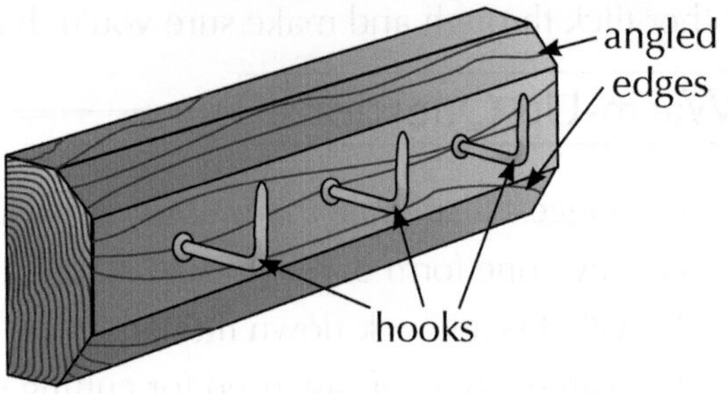

Figure 4

   a)   State a type of saw that could be used to cut
        the length of oak to the right size.

        ....
        [1 mark]

   b)   Once sawn, the ends of the key rack are sanded with abrasive paper to smooth them down.
        Give **one** benefit of using a sanding block when sanding.

        ....

        ....
        [1 mark]

   c)   The carpenter presses a bradawl into the wood where a hole is to be made for each hook.
        Explain why the carpenter does this before the holes are drilled.

        ....

        ....
        [2 marks]

   d)   Name a hand tool that could be used to produce the angled edges of the front of the key rack.

        ....
        [1 mark]

2   A rivet can be used to permanently join sheets of metal.

    Use sketches and/or notes to explain how a rivet can be used for this purpose.

                                                                                        [4 marks]

Shaping Materials — Power and Machine Tools

Finally — the bit you've been waiting for — using power tools to cut and shape stuff.
It doesn't get much better than this. Before you get stuck in though, a little safety announcement...

Safety is **Really Important**

There's more about safety on page 151.

Power tools are hand-held motorised tools. You need to use them safely...

1) Before using power tools, do a visual check for any loose connections and run your hand along the lead to check for any cuts in the insulation (when it's not plugged in, of course). Check that the blade or drill bit or whatever is attached correctly and tightly.

2) You can use an RCD (Residual Current Device) to help prevent electric shocks. The power tool plugs into the RCD, which you plug into the socket. If you accidentally cut through the lead of the power tool, the RCD cuts off the electricity supply straight away.

3) Wear a mask or fit an extraction hose if the tool's going to produce a lot of dust. Always wear safety glasses and make sure clothing can't get caught.

4) Clamp your work down firmly so it can't slip or move.

5) Make sure you know where the stop buttons are before you start.

6) When you've finished, make sure the tool has stopped moving before you put it down.

Power Tools can be Used to **Cut**, **Shape** and **Smooth** Materials

Like with hand tools, there's a huge range of power tools for shaping materials. Here are a few examples:

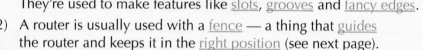

Routers

1) Hand-held routers have a spinning cutting tool that cuts away wood. They're used to make features like slots, grooves and fancy edges.

2) A router is usually used with a fence — a thing that guides the router and keeps it in the right position (see next page).

3) You can get different cutting tools to make different shapes.

You met CNC routers on page 5.

Planers

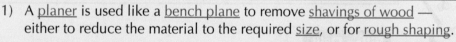

1) A planer is used like a bench plane to remove shavings of wood — either to reduce the material to the required size, or for rough shaping.

2) The advantage of a power planer is that it takes much less effort and is much faster — but it's not as accurate as a bench plane.

3) Before you start, check the blades are sharp and replace them if not.

4) Don't start with the cutting tool on the wood — rest the flat front base on the wood and start to push it forward when it reaches full speed.

5) Make sure you're well-balanced and use two hands to hold it — one on the front handle and one on the trigger switch.

Jigsaws

1) A jigsaw has interchangeable blades and variable speeds.

2) You can make straight or curved cuts in all materials, but it's quite slow. A fence helps you make straight cuts.

3) Make sure the blade is secured tightly and the correct type of blade is installed for the material. The teeth should face the front of the saw, and you should push it forwards (away from you). Don't start cutting until the blade is at full speed.

Sanders

1) Electric power sanders can be used to smooth wood.

2) They work by moving abrasive paper at high speeds. This is a lot quicker and easier than sanding by hand.

3) Different types of power sanders work in different ways. For example, belt sanders have a loop of abrasive paper which rotates at high speeds.

4) Most sanders have a dust bag or an extraction hose to remove the dust that they produce.

Shaping Materials — Power and Machine Tools

Machine Tools are Great for Bigger Jobs

1) Machine tools generally do the same jobs as power tools — but they're often more accurate and better suited to bigger tasks.

2) Machine tools are usually stationary, often bolted to the workbench or the floor. They're usually attached to a dust extractor too for when wood is being machined. These tools can be used to process large quantities of material accurately and quickly.

Machine Tools can be used to Cut, Drill and Sand

Sanding Disc

1) A sanding disc spins a disc of abrasive paper which the material's pushed against.

2) It's used for accurately removing material to a line.

3) Different types of abrasive paper can be used on wood, metal, and plastics.

Saw Bench

The saw bench has a circular blade and is used to cut wood and boards like plywood to size. It makes straight cuts only.

Saw bench

There are more machine tools on the next page.

Band Saw

The band saw has a blade in a long flexible loop and is normally used to cut wood, but special blades can be bought for use on plastics and softer metals. The blades come in different widths and can be used for straight or curved cuts.

Pillar Drill

A pillar drill (or pedestal drill) is used to make round holes. They can be used on all kinds of materials, depending on the drill bit used (see page 82).

Fences are Attached to Tools to Improve Accuracy

Router

Fence attached to router

1) Fences are attached to power and machine tools to help guide either the tool or the material.

2) In the example on the right, the fence runs along the edge of the material, keeping the router the same distance from the edge at all times. This keeps the cut straight.

3) Fences can be set up to repeat the same action over and over again — ensuring consistency between the products in a batch (see p.47).

It's important to be as accurate as possible when using tools...

When using tools to cut a material, you need to check the part you've made is within tolerance (see p.49).

Shaping Techniques

There are loads of <u>techniques</u> for getting materials in the <u>right shape</u>. The best one for the job depends on the <u>shape</u> you're <u>aiming for</u> and the <u>type of material</u> you're shaping. Here are some to get you started...

Milling Machines Remove Thin Layers of Material

1) <u>Milling machines</u> remove material one <u>thin layer at a time</u> to produce the required size or shape.

2) These machines can be set to a particular <u>cutting speed</u> and <u>depth</u>.

3) They can also be used to make a surface <u>absolutely flat</u> and can produce a very <u>accurate finish</u>.

Lathes Are For Turning

Turning means shaping a material while it rotates.

1) <u>Lathes</u> come in two types — <u>wood lathes</u> and <u>engineers' lathes</u> (also known as a centre lathe, for working metal). They're used to 'turn' materials, to make objects like these.

2) A material is <u>held</u> and <u>rotated</u> by the lathe, while a <u>tool</u> or <u>bit</u> is pressed onto the material to cut it.

3) Make sure the material is held <u>tightly</u> and <u>straight</u>. If you're turning a <u>long</u> piece of wood or metal, it needs to be held securely at <u>both ends</u> to stop it <u>wobbling</u>.

> Both milling machines and lathes can be used in <u>CAM</u>. They are <u>subtractive</u> CAM processes — material is <u>removed</u> from a larger piece of material to make the product.

Take a look at page 4 for how milling machines are used with CAM.

3D Printing is an Additive CAM Process

1) <u>Additive</u> CAM processes involve <u>adding material</u> to build up the product rather than by removing it.

2) <u>3D Printing</u> is <u>additive</u>. It works by printing <u>layers</u> of molten plastic, powder or wax until the full 3D shape has been formed.

3D printing is often used to make prototypes (p.5) and models (p.137).

3) You can <u>design</u> your ideas on screen using various <u>software</u> packages (this is an example of <u>CAD</u> — see p.4), then use a <u>3D printer</u> to convert your design into a <u>3D model</u>.

Metal Sheets can be Pressed into Shape

1) <u>Press forming</u> involves pressing metal sheets between <u>two moulds</u> with a <u>large amount</u> of <u>force</u>.

2) The <u>metal sheets</u> to be pressed are often <u>annealed</u>, as this means they're more <u>malleable</u> (see page 18).

Here's how it's done:

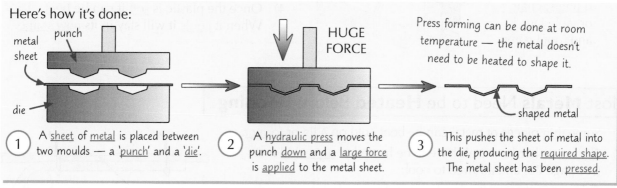

Press forming can be done at room temperature — the metal doesn't need to be heated to shape it.

① A <u>sheet</u> of <u>metal</u> is placed between two moulds — a '<u>punch</u>' and a '<u>die</u>'.

② A <u>hydraulic press</u> moves the punch <u>down</u> and a <u>large force</u> is <u>applied</u> to the metal sheet.

③ This pushes the sheet of metal into the die, producing the <u>required shape</u>. The metal sheet has been <u>pressed</u>.

3) <u>Similar</u> machines such as <u>punches</u> and <u>die cutters</u> can <u>cut out</u> parts from a sheet of material (e.g. a metal or plastic). The machine applies <u>shear forces</u> (see p.45) to deform the material in this way.

Casting is Used to Shape Materials

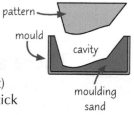

1) The process of casting involves <u>molten material</u> being poured into a <u>hollow mould</u>. The material is then left to <u>cool and solidify</u> before being <u>separated</u> from the mould.

2) Before casting a mould must be made. A <u>pattern</u> (an exact replica of the item to be cast) is used to create a <u>cavity</u> in a box filled with moulding sand (special sand that doesn't stick to the material being cast). This cavity will be <u>filled with material</u> during casting.

3) Moulds are often made in <u>two halves</u> so they can be split open to release the cast material.

Section Five — Wood, Metals and Polymers

Shaping Techniques

Die Casting Can be Done on Metals and Plastics

1) Die casting is used to mould <u>metals</u> and <u>thermoforming plastics</u>.

2) The material is <u>melted</u> and poured into a <u>mould</u> (the 'die') which is in the shape of the product.

3) It is then allowed to <u>cool</u>. Once it has <u>solidified</u> it can be removed from the mould.

4) Some plastic resins can be <u>cold-poured</u> into moulds (without heating). They <u>harden</u> or <u>set</u> through a <u>chemical reaction</u>.

5) When the product is <u>removed</u> from the mould it may need to be <u>trimmed</u> to remove any <u>excess material</u>.

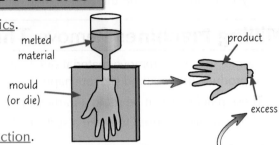

There are a Few Ways That Materials can be Bent

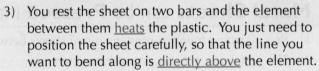

1) You can use a <u>SHEET METAL FOLDER</u> to shape <u>sheet metals</u> such as aluminium and tin.

2) The outline of the product, e.g. a box, is marked out and cut from a <u>flat</u> sheet of metal.

3) You <u>feed the metal in</u> flat, make one fold then move the material through for the next fold.

4) Corners can then be <u>joined</u> using rivets (see page 79), or by soldering, brazing, etc. (see page 90).

1) <u>Wood</u> can be bent by <u>LAMINATING</u> it. There's more on lamination on p.46.

2) <u>Thin strips</u> of wood are <u>glued together</u> and held in a <u>jig</u>, which keeps them <u>bent</u> into the <u>right shape</u>. Once the glue has <u>dried</u>, the strips of wood can be taken out of the jig, and <u>stay bent</u>.

3) Things that could be made this way include <u>rocking chair runners</u>, chair or <u>table legs</u> and <u>roof beams</u>.

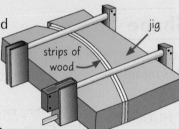

1) <u>LINE BENDING</u> is ideal for use with <u>acrylic sheets</u>, e.g. for making picture frames, pencil holders, etc.

2) It can be done manually or with a <u>line bender</u> or <u>strip heater</u>.

3) You rest the sheet on two bars and the element between them <u>heats</u> the plastic. You just need to position the sheet carefully, so that the line you want to bend along is <u>directly above</u> the element.

4) Once the plastic is <u>soft</u> it can be <u>bent</u>. When it <u>cools</u> it will stay in its <u>new shape</u>.

Most Metals Need to be Heated Before Bending

1) Some <u>thin</u> pieces of metal can be bent cold on a <u>jig</u> or <u>former</u>.

2) <u>Thicker</u> or harder metals have to be heated or <u>annealed</u> first (see page 76) and allowed to cool.

3) This makes them soft enough to bend easily, but the annealing process might have to be repeated as bending makes them go <u>hard</u> again — this is known as '<u>work hardening</u>'.

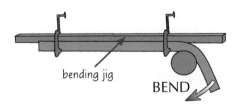

There are several different shaping techniques to know about...

If you're asked to <u>outline</u> how to do something, make sure you include enough detail about the particular process you'd use. For example, don't just say, 'using a sheet metal folder' — you need to be able to describe the <u>different steps</u> of the process involved to get all the available marks.

Moulding and Joining

There are plenty of ways to <u>mould</u> materials and <u>join</u> them <u>together</u> — especially <u>plastics</u> and <u>metals</u>...

Air is Sucked Out In **Vacuum Forming**

1) A mould is put onto the <u>vacuum bed</u>.

2) <u>Thermoforming plastic</u> sheets or films
(e.g. polypropylene or HIPS) are <u>clamped above</u>
the <u>vacuum bed</u> and <u>heated</u> until they go <u>soft</u>.

3) The vacuum bed is <u>lifted close</u> to the heated plastic.

4) The air is <u>sucked</u> out from under the plastic,
creating a <u>vacuum</u>. The air pressure from outside
the mould then <u>forces the plastic onto the mould</u>.

5) The moulded plastic is <u>cooled</u> and the vacuum bed <u>lowered</u>.
The <u>cold</u> plastic is <u>rigid</u> so holds its <u>new shape</u>.

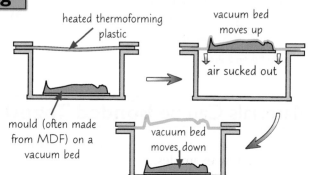

heated thermoforming plastic

vacuum bed moves up

air sucked out

mould (often made from MDF) on a vacuum bed

vacuum bed moves down

Blow Moulding... Well... Blows Air In

1) A tube of <u>softened plastic</u> is
inserted into a <u>solid mould</u>.

2) <u>Air</u> is then injected which
forces the plastic to <u>expand</u>
to the <u>shape</u> of the <u>mould</u>.

3) This method is often used
to produce <u>bottles</u> and
<u>containers</u>.

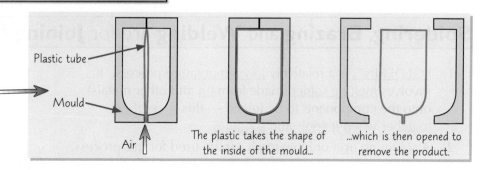

Plastic tube

Mould

Air

The plastic takes the shape of the inside of the mould...

...which is then opened to remove the product.

Injection Moulding Uses **Pressure** to Mould Plastics

1) Injection moulding is similar to casting (see
p.87), but the molten material is forced
into a <u>closed mould</u> under <u>pressure</u>.

2) The moulds are often made from <u>tool steel</u> —
it's very hard but it's also quite expensive.
The plastic is often melted using <u>built-in heaters</u>.

3) It can be used to make things like
<u>plastic buckets</u> and <u>watering cans</u>.

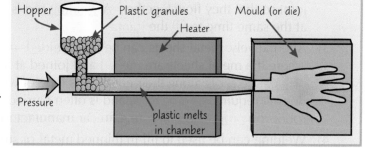

Hopper

Plastic granules

Mould (or die)

Heater

Pressure

plastic melts in chamber

In both vacuum forming and injection moulding the moulds must
have rounded corners and be slightly tapered (sloped) at the sides
— so that the finished product can be released from the mould.

Extrusion Produces **Long, Continuous Strips**

1) Extrusion is similar to injection moulding. It's used
for <u>thermoforming plastics</u> and some <u>metals</u>.

2) The material is <u>melted</u> and forced
under <u>pressure</u> through a <u>die</u>.

3) It produces long, <u>continuous</u> strips of the moulding
exactly the same shape as the exit hole.
It's used for products like <u>plastic-covered wire</u>,
and <u>plastic and aluminium window frames</u>.

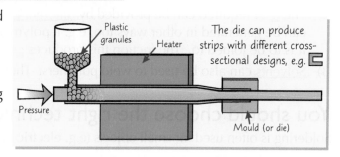

Plastic granules

Heater

The die can produce strips with different cross-sectional designs, e.g.

Pressure

Mould (or die)

Moulding and Joining

Drape Forming Moulds Softened Thermoforming Plastics

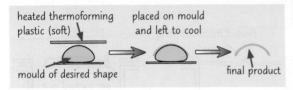

heated thermoforming plastic (soft) placed on mould and left to cool

mould of desired shape final product

1) A sheet of thermoforming plastic is <u>heated</u> until it <u>softens</u>.
2) The softened sheet takes the <u>shape</u> of the <u>mould</u> that it is placed on. (It's a bit like putting a table cloth on a table — it takes the shape of the table.)
3) Once the plastic has <u>cooled</u> it will <u>stay that shape</u>.

Materials Can be Bonded Together with Glue

There are <u>all kinds</u> of <u>adhesives</u> (glues) — which one you use depends on the <u>material</u> you're <u>joining</u>. Here are a few examples:

1) <u>Polyvinyl acetate (PVA) glue</u> is used with <u>wood</u>, <u>paper</u> and <u>card</u>. It <u>dries slowly</u>.
2) <u>Glue guns</u> use melted plastic to join <u>woods</u> and <u>fabrics</u>. It's especially good for <u>modelling</u>.
3) <u>Solvent cement</u> is used to join some <u>plastics</u>. The joint needs to be <u>clamped</u> whilst it sets.
4) <u>Epoxy resin</u> and <u>super glue</u> can be used on <u>most materials</u>. They <u>set quickly</u> but <u>aren't cheap</u>.

Soldering, Brazing and Welding are for Joining Metal

There's more about soldering on page 117.

1) <u>SOLDERING</u> is a relatively <u>low temperature</u> process. It involves <u>melting solder</u> (made from <u>tin</u> and other metals) onto the components to be joined — this <u>sticks them together</u> when it <u>cools</u> and <u>solidifies</u>.
2) A <u>soldering iron</u> or <u>blow torch</u> can be used for this process.

1) <u>WELDING</u> is by far the <u>strongest</u> method of joining metal.
2) It uses a <u>very high temperature</u> from a <u>gas</u> (oxyacetylene) <u>torch</u>, or an <u>electric-arc welder</u> to <u>melt</u> the edges of the joint so that they flow together. A <u>welding rod</u> is <u>melted</u> at the same time to <u>fill</u> the <u>joint</u>.
3) Alternatively, metal sheets can be <u>spot welded</u> — this is where the metal sheets are <u>melted</u> and joined at a number of <u>separate spots</u> along their edge. This type of weld doesn't require <u>welding rods</u>, and is often carried out by <u>robots</u> on <u>production lines</u> (e.g. in car manufacturing).
4) Welding can be used to fill in thinned metal or small <u>gaps</u>.

1) <u>BRAZING</u> is a <u>higher temperature</u> process that often uses <u>brass spelter</u> as the joining material. It's much <u>stronger than soldering</u>.
2) Either a <u>gas brazing torch</u>, a <u>blow torch</u>, or a brazing attachment for an <u>electric-arc welder</u> is used to heat the joint.

Welding masks are worn to protect your eyes from the bright light and UV radiation from the welding arc. It also protects your face from heat and sparks.

welding mask welding rod

Polymers can be Welded Too

1) Polymers can also be welded together using <u>high temperatures</u>. Like with metal welding, the process <u>melts</u> the edges of the joint so that they flow together. <u>Welding rods</u> made from <u>thermoforming plastic</u> can be used to <u>melt plastic</u> into the join. *Pressure is applied to the joined parts during cooling.*
2) The <u>heat</u> required can be provided by <u>different tools</u> such as <u>hot gas welding guns</u>, <u>lasers</u>, etc. Heat can be generated in other ways too, e.g. a polymer can be <u>vibrated</u> at high speeds <u>against</u> the polymer it is to be joined to. The <u>friction</u> this produces <u>generates heat</u>, which welds the parts together.
3) <u>Solvents</u> can also be used to weld polymers. They chemically 'melt' the polymers to join them together.

You should choose the right technique for the job in hand...

Soldering is often used for small objects (e.g. electric circuits), whilst welding is used for larger items (e.g. cars).

Treatments and Finishes

There are loads of different <u>treatments</u> and <u>finishes</u> that you can apply to <u>timber</u>, <u>metals</u> and <u>polymers</u>.

Prepare Wood Before Applying Finishes

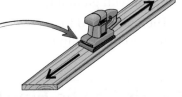

1) Before applying <u>any finish</u>, timber needs to be <u>sanded</u> to a smooth finish using <u>abrasive paper</u> (see page 81). Sand in the direction of the grain.

2) It's often a good idea to apply a <u>sanding sealer</u> before sanding timber. Otherwise, the first coat of <u>varnish</u> or <u>wax</u> can cause loose wood <u>fibres</u> to <u>stick up</u>, and the timber will have to be <u>sanded again</u>.

3) Before <u>painting</u>, <u>bare timber</u> needs to be <u>primed</u>. Priming does three useful things:
   - <u>Fills</u> the <u>grain</u> of the timber to give the paint on top a <u>smooth finish</u>.
   - Helps <u>seal</u> the timber, so paint won't <u>soak in</u> and need to be re-applied.
   - Helps the paint to <u>stick better</u> to the timber.

There Are Many Types of Paint

Paints are used to <u>protect timber</u>, when you want to <u>cover up the grain</u> and change its <u>colour</u>. Between <u>each layer</u> of paint you <u>sand lightly</u> in the direction of the grain.

Undercoat	<u>Undercoat</u> is the first coat of paint (on top of a primer if you've used one, or on old finishes if you're reusing timber). It <u>covers up</u> any previous <u>colours</u> more <u>cheaply</u> than applying extra layers of the final 'topcoat', which is usually a more expensive paint. It helps later layers of <u>paint</u> to <u>stick</u>.

Top coats	1) <u>Gloss paints</u> are <u>hard-wearing</u> and <u>waterproof</u> and come in lots of <u>colours</u>. They're shiny and used for things like interior woodwork. You apply them with a <u>brush</u> or <u>roller</u>, painting in the <u>direction of the grain</u>. 2) <u>Polyurethane paints</u> are even <u>tougher</u> than gloss paints. They're used for things like <u>children's toys</u>. They're often <u>sprayed</u> on for a <u>smooth finish</u>.

So, deep breath: sand, prime, sand, undercoat, sand, paint, sand, paint.

Varnish Lets You See the Grain

Many woods (especially hardwoods) have an <u>attractive grain</u> — so you might want to use <u>clear varnish</u> rather than paint as a surface finish. For example, oak furniture is often varnished.

1) Varnish can be <u>coloured</u> or <u>clear</u>, and either <u>gloss</u> (shiny), <u>matt</u> (dull) or <u>satin</u> (in between).

2) <u>Yacht varnish</u> seals the wood and makes it <u>waterproof</u>. It's pretty <u>flexible</u>, so it <u>doesn't crack</u> if the wood <u>moves</u>. So it's good for outdoor uses, e.g. doors and window frames (and yachts).

3) <u>Polyurethane varnish</u> is best for <u>interior</u> uses, e.g. stairs and skirting boards. It's very <u>hard-wearing</u>.

4) For best results, you need to apply <u>two or three coats</u>, <u>sanding</u> lightly <u>between</u> each coat.

Wood can be Treated in a Process Called Tanalising

Woods that will be used <u>outdoors</u> are often treated with a <u>wood preservative</u>. This helps to prevent <u>insect attacks</u> and <u>decay</u> of the wood, meaning it will <u>last longer</u>. Here's how it's done:

Putting the wood in a vacuum removes air from the cells of the wood — this means the preservative can go there instead.

1) Seasoned timber is placed into a <u>treatment tank</u>.
2) The <u>air</u> is <u>removed</u> from the tank, creating a <u>vacuum</u>.
3) The tank is <u>flooded</u> with the <u>preservative</u>.
4) <u>Pressure</u> is applied which <u>forces</u> the preservative <u>deep</u> into wood.
5) <u>Excess preservative</u> is <u>removed</u> and the timber is left to <u>dry</u>.

Tanalised® wood is used for things like <u>outdoor playgrounds</u>, <u>fences</u>, <u>telegraph poles</u>, etc.

Treatments and Finishes

Metals Need To be Prepared Before Finishing

1) Metals have to be smoothed first. You can do this by filing. Next, you rub the metal with gradually finer grades of abrasive paper and wet and dry paper.

2) Then you need to remove grease from the surface (by rubbing with a cloth soaked in paraffin or degreaser) — to make sure the finishes will stick properly.

Metals Can be Finished for Protection and Looks

1) Some metals don't need a finish, but some can corrode without suitable protection.

2) Corrosion is where metals react with substances, become oxidised and are gradually destroyed.

3) Finishes can also improve the appearance of your product.

Dip coating
- A metal is heated evenly in an oven and then plunged into fluidised powder (very fine plastic powder that's made to act like a liquid by passing gas through it) for a few seconds.
- The metal, with this thin coating of plastic, is then put back in the oven and the plastic fuses (joins completely) to the surface.
- It gives a soft, smooth finish so it's used for things like wire racks and tool handles.

Powder Coating
- Plastic powder is sprayed onto the metal using an electrostatic gun.
- The spray gun gives the powder an electrical charge. The particles repel all the others, since they've all got the same charge, so you get a very fine, even spray.
- The object to be coated is given an opposite charge to the gun. This attracts the fine spray.
- After spraying, the object is heated in an oven so that it undergoes a chemical change and sets hard.
- This method gives an even coat and there's hardly any waste.

Galvanising
- Iron rusts when it comes into contact with both oxygen and water, which are present in air.
- Coating the iron with a barrier to keep out the water and oxygen can stop it rusting (oxidising). Another way is to place a more reactive metal such as zinc with the iron — water and oxygen then react with the zinc instead of with the iron.
- Coating an object with zinc is known as galvanisation. The zinc layer protects the iron, but also if it's scratched, the zinc around the site of the scratch reacts first rather than the iron.

Polymers are Self-Finishing But You Can Decorate Them

1) Plastics don't need protective surface finishes because they're very resistant to corrosion and decay.

2) But for a nice appearance, you can use wet and dry paper to remove scratches, and follow that up with a mild abrasive polish or anti-static cream. Or, you could use a buffing machine.

3) Plastics can have decorative vinyl decals applied to them. Sheets of vinyl are cut to a design. The vinyl has an adhesive coating on the back so that it can be transferred to a plastic surface.

4) Designs can also be printed directly onto plastics. This can be done in a number of ways, for example, offset lithography (see p.68).

Finishes are a useful way of improving the aesthetics of a product...
You may need to describe how to prepare a surface for finishing or how to apply a finish, so make sure you can.

Warm-Up and Worked Exam Questions

Well that's Woods, Metals and Polymers all wrapped up. As always, try the warm-up questions to start, look over the worked exam questions and then have a go at the exam questions on the next page.

Warm-Up Questions

1) Describe what a jigsaw can be used for.
2) Name a piece of equipment that can be used to bend a sheet of aluminium.
3) Describe the process of drape forming.
4) Name the process of coating metal in a thin layer of zinc.

Worked Exam Questions

1 Power tools are hand-held and motorised.

a) Give **three** safety precautions that need to be taken when working with power tools.

1. Do a visual check for any loose connections.

2. Check for any cuts in insulation along the lead.

3. Check the blade or drill bit is attached correctly.

Any other sensible precaution would be fine here too — for example, wear safety glasses. [3 marks]

b) Machine tools generally do the same jobs as power tools, but are not hand-held. Give **one** advantage of machine tools compared to power tools.

Machine tools are often more accurate than power tools.

[1 mark]

2 The oak dining table shown in **Figure 1** has been finished with a varnish.

a) Describe one way in which the oak could have been prepared before varnishing.

A sanding sealer should be applied to the oak. The oak should than be sanded in the direction of the grain using abrasive paper.

[2 marks]

Figure 1

b) Suggest why varnish has been chosen rather than paint to coat the surface of the table.

Oak grain markings are attractive. They can be seen beneath a varnish but not beneath paint.

[1 mark]

c) The table needs a finish that protects it from knocks and scratches. Suggest a suitable type of varnish that could be used, and give a reason for your answer.

Polyurethane varnish, because it is very hard-wearing.

[2 marks]

Exam Questions

1   The body of the model car shown in **Figure 2** is made of metal, and has been produced using die casting.

Use sketches and/or notes to give a detailed description of the die casting process for this product.

Figure 2

[4 marks]

2   The products shown in **Figure 3** are all made using a mould.

Choose **one** product in **Figure 3** and name the moulding process that is used to manufacture it. Use sketches and/or notes to give a detailed description of the moulding process of this product.

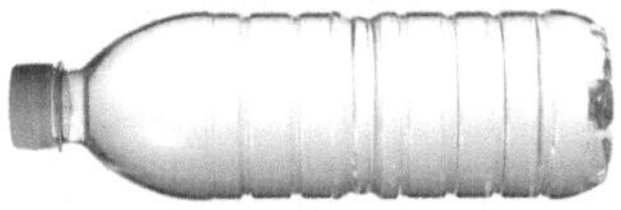

| Plastic bottle | Plastic guttering |

Figure 3

Name of product/component: ....

Moulding process: ....

[5 marks]

Revision Questions for Section Five

That's just about it for <u>Section Five</u> — so now's a good time to <u>test</u> your <u>knowledge</u> with some questions.
- Try these questions and <u>tick off each one</u> when you <u>get it right</u>.
- When you've done <u>all the questions</u> for a topic and are <u>completely happy</u> with it, tick off the topic.

Uses of Wood, Metals and Polymers (p.75-76) ☐

1) Bill is making a traditional wooden toy using beech.
   Give two properties of beech that make it a suitable material to make the toy from. ☐

2) Some cooking utensils are made from stainless steel.
   Give two properties of stainless steel that makes it suitable for use in cooking utensils. ☐

3) Flat pack furniture is often made from MDF.
   State two properties of MDF that make it a good material for flat pack furniture. ☐

4) a) What is the name of the process used to soften metal?
   b) Outline what happens during this process. ☐

Stock Forms and Standard Components (p.77-80) ☐

5) Give three example of timber stock forms. ☐

6) Name a material that screws are often made from. ☐

7) What are machine screws commonly used for? ☐

8) Jack is making a door for a small kitchen cupboard. Suggest what type of hinge he could use. ☐

9) Give one advantage and one disadvantage of using knock-down fittings. ☐

Shaping Materials — Hand, Machine and Power Tools (p.81-82 & p.85-86) ☐

10) Which of the following types of saw is best for cutting metals?
    a) rip saw   b) tenon saw   c) hacksaw   d) coping saw ☐

11) Rob wants to remove a thin layer of wood from the bottom of a door he is fitting.
    Suggest a hand tool that he could use to do this. ☐

12) Give two pieces of safety advice you would give to someone using a rip saw for the first time. ☐

13) Susan is making a door. Suggest the power tool that she should use to:
    a) cut the door down to the right height.
    b) make a groove 10 cm in from the edge of the door. ☐

Shaping Techniques (p.87-88) ☐

14) Describe what a milling machine is used for. ☐

15) a) Why is press forming often carried out on metals that have been annealed?
    b) Outline the process of press forming. ☐

16) Name a process that could be used to fold a piece of acrylic. ☐

Moulding, Joining, Treatments and Finishes (p.89-92) ☐

17) Describe the process of vacuum forming. ☐

18) What is solder made from? ☐

19) What is the process used to treat timber to protect it from insect attacks and decay outdoors?
    a) priming   b) tanalising   c) undercoating   d) galvanising ☐

Fabrics and Their Properties

Different materials have different <u>properties</u>. Flick back to <u>Section Two</u> if you need to refresh your memory.

The **Properties** of a Fabric Decide What It will be **Used For**

When deciding what fabric to use for a product, it's important to think about its <u>properties</u>.

Sportswear

Sportswear is often made from <u>polyester</u>, <u>elastane</u> or <u>polyamide</u> fibres. It needs to be <u>light</u> and <u>breathable</u>, and have other properties...

1) Polyester is <u>resistant to abrasion</u> — this means it can stand up to being used <u>intensively</u> (e.g. whilst exercising) again and again. It's also <u>strong</u> even when wet and <u>dries very quickly</u> — useful when being worn during exercise and when sweating.

2) Elastane is <u>really stretchy</u> so it's super flexible — great for when you're exercising. It's <u>not absorbent</u> and is usually used in <u>combination</u> with other <u>fibres</u> which are often able to <u>wick moisture</u> away from the body. Elastane is often used in <u>swimwear</u> and <u>cycling clothing</u>.

3) Polyamide fabric is made from a synthetic fibre. It's an <u>insulator</u> so can be very <u>warm</u> — useful for outdoor sportswear such as <u>ski-jackets</u>. It's also <u>easy to wash</u> (a very desirable quality for sweaty sportswear).

Furnishings

Furnishings are often made of <u>cotton</u>, <u>acrylic</u> or <u>wool</u> fibres. They need to be <u>soft</u> and <u>comfortable</u>, but also <u>durable</u>.

1) Cotton is used to make furnishings like cushions and curtains. It's <u>resistant to abrasion</u> — so it won't wear thin very easily, even when being sat on or handled every day.

2) Acrylic is <u>soft</u> and is quite <u>warm</u>, so will help to make your home nice and <u>cosy</u>. It's also <u>resistant to fading</u>, so will stay looking good for years. A <u>blend</u> of acrylic and other fibres is often used to <u>cover furniture</u> like sofas and chairs.

3) Wool is <u>strong and warm</u> — it's often used to make rugs and blankets.

In the 1900s, some swimsuits were made of wool. Hot, heavy and super absorbent — perfect for a sunny day on the beach...

Treatments can **Change** the **Properties** of Fabrics

1) <u>Chemical</u> treatments are applied to fabrics to <u>change their functional properties</u>. They are applied <u>during manufacture</u> at different stages of the production process. Here are a few for you to take a look at...

Flame retardance

1) <u>Flame retardant</u> treatments make fabrics <u>less likely to catch fire</u>.

2) They're often used on <u>flammable</u> fibres like <u>cotton</u>, and for specific products such as:

- Work-wear for <u>welders</u> — e.g. overalls. Welders work with <u>flames</u> and <u>hot metal</u> so their clothing needs to be protected.
- <u>Racing drivers</u>' overalls — their clothes need to be flame retardant in case they <u>crash</u>.
- <u>Night clothes</u> (especially children's pyjamas). <u>Young</u> children's skin is more susceptible to <u>burns</u>, and they're <u>less aware</u> of the <u>risks</u> of <u>fire</u>, so it's important that their clothing is protected.
- Fabric for <u>soft furnishings</u> — to make them meet <u>fire safety requirements</u>.

3) Using a fire retardant treatment on fabrics like cotton makes the fabric <u>slightly stiffer</u>, but the fabric is still <u>soft</u> and <u>cheap to produce</u>.

4) Some flame retardant treatments can be <u>washed out</u>, so care is needed when washing.

Chemicals can also be used to change the aesthetic properties of a fabric — check out pages 104-107.

Fabrics and Their Properties

Stain Protection

1) Fabrics can be made stain resistant by treating it with a mixture of silicone and fluorine compounds, or a Teflon® coating (this is also used on non-stick frying pans).
2) These treatments stop grease and dirt from penetrating the fabric.
3) Stain resistant treatments are used a lot on carpets and upholstery.
4) More recently, nanoparticles have been used in "self-cleaning fabrics" to give improved stain resistance (see p.38).

Rot Proofing

1) Mildew is a type of fungus — it can grow on fabrics made from natural fibres if they're kept in damp conditions.
2) If this is left untreated it can break down the fabric and cause it to rot.
3) Mildew growth can be reduced by applying a waterproof treatment such as PVC.

Some fabrics, like acrylic, are resistant to mildew.

Water-resistant Finishes

1) Chemicals (e.g. silicones) can be applied to the surface of fabrics to stop water droplets passing through.
2) These finishes don't make the fabric waterproof — if the surface becomes saturated (completely covered in water) the water will leak through.
3) Fabrics with a water-resistant finish can be washed and dry-cleaned without affecting the performance of the finish.
4) Nylon is often given a water-resistant finish and used to make coats and tents.

Water forms beads on the surface of the fabric.

2) The properties of a fabric aren't only changed by chemical treatments though — how the fabric has been constructed can also have an impact...

Laminated Fabrics are Constructed from Different Layers

1) Laminated fabrics are made up of two or more different layers that are stuck together.
2) Lamination can add useful properties, e.g. a layer of foam could be used to make a fabric more insulating. It can also be used with delicate fabrics — to add strength from an under layer, or protection from an outer layer.

For other ways of adding strength to fabric see page 46.

3) Waterproof coats are made from laminated fabrics. They have to be warm, breathable and waterproof, so are often made of a layer of natural or synthetic fibres that is then laminated with a waterproof layer such as rubber or PVC.
4) Some laminated fabrics in waterproof coats also use a membrane layer that allows sweat to escape but doesn't let rain in. This helps to keep the wearer warm and dry.

Water-resistant isn't the same as waterproof — it could still leak...

So, different fabrics have properties that make them useful for different things. And properties can be added to a fabric to make it safer or more useful. Wonderful stuff. Make sure you know it from top to bottom.

Standard Components and Tools

Next up, you need to know about <u>how fabric is sold</u> and a few <u>standard components</u> you can <u>use</u> with <u>fabrics</u>.

Fabric is **Sold** in **Standard Widths**

When you're buying fabric you normally just choose the <u>length</u> that you want — the width is determined by the <u>width of the roll</u> that you buy it from.

1) Fabric comes on <u>rolls</u> of standard <u>widths</u>, e.g. 90 cm, 115 cm and 150 cm. So you only need to specify the <u>length</u> you require.

2) Fabrics also come in different <u>weights</u>. The weight of a fabric depends on how <u>tightly</u> the fabric is knitted or woven, and on the <u>thickness</u> of the <u>yarn</u> used (i.e. 1-ply, 2-ply etc.).

EXAMPLE:

A 4.5 m length of fabric is cut from a 115 cm-wide roll. Calculate the area of this piece of fabric.

First convert the lengths to the same unit (e.g. metres).
1 m = 100 cm, so 115 cm = 115 ÷ 100 = 1.15 m

Then calculate the area by multiplying length by width.
= 4.5 m x 1.15 m
= 5.18 m² (to 2 decimal places)

Standard Components include **Fastenings**

<u>Standard components</u> are the bits and pieces that you use in <u>addition</u> to the fabric to make a <u>textile product</u>. They can be <u>functional</u>, e.g. fastenings like zips to close your coat, or <u>decorative</u>, e.g. additional buttons.

Zips

1) They can be made out of <u>plastic</u> or <u>metal</u>, and can be <u>big</u> and <u>bulky</u> or <u>small</u> and <u>concealed</u> (hidden) in your textile design.

2) Some zips are <u>fixed</u> at <u>one end</u> (e.g. on handbags). Zips on <u>jackets</u> are <u>not fixed</u> (so you can get the jacket on and off).

3) Zips with <u>two sliders</u> can be opened in two directions — the ends can be fixed (e.g. on suitcases) or open (e.g. on some jackets).

Advantages

1) They're a <u>secure</u> fastening — they close the product fully with no gaps.
2) They're quick to <u>do up</u> and <u>undo</u>.
3) They're <u>quick</u> and <u>simple</u> to attach.
4) They <u>lay flat</u> and don't add <u>bulk</u>.
5) They're <u>hard-wearing</u> and can be <u>washed</u>.
6) Colours can <u>match</u> or <u>contrast</u> with the fabric.

A **disadvantage** is that they can <u>snag</u> delicate fabrics.

Velcro®

1) Velcro® comes in two halves — a rough tape and a smooth tape. Nylon <u>hooks</u> on the rough half attach to soft <u>loops</u> on the smooth half.

2) Velcro® requires a large amount of force to open it — so it needs to be <u>firmly attached</u> to the fabric and isn't suitable for using with <u>delicate fabrics</u>.

Advantages

1) <u>Safe</u> and <u>soft</u> (good for <u>children's products</u>).
2) It can be <u>machine washed</u>.
3) It's <u>hard-wearing</u>.

Disadvantages

1) <u>Hooks</u> collect <u>fibres</u> over time and become <u>less sticky</u>.
2) Not very <u>decorative</u>.

Toggles and Buttons

1) These are sewn on and require a <u>buttonhole</u> or a <u>loop</u> to fasten to.

2) They can be made of <u>any hard</u> material — <u>plastic</u>, <u>metal</u>, <u>wood</u>, and even <u>glass</u>.

Advantages

1) They're easy to <u>attach</u> and <u>replace</u>.
2) Colours can <u>match</u> or <u>contrast</u> with the fabric. (Buttons can be <u>fabric-covered</u> to blend in.)

Disadvantages

1) They can <u>fall off</u> — a <u>choking</u> hazard on children's products.
2) They can be <u>damaged</u> in the <u>wash</u>.

Press studs (or poppers)

These can be used to fasten items that need to be opened and closed <u>quickly</u>. They can be made of <u>metal</u> or <u>plastic</u> and come in different <u>sizes</u> depending on the <u>strength</u> of fastening you require (<u>bigger metal</u> press studs are <u>harder</u> to open than <u>small plastic</u> press studs). And they're <u>not very decorative</u>.

Standard Components and Tools

You also need to know about <u>tools</u> that can be used when <u>making textiles products</u>...

Use the **Right Tools** for Each **Task**

For Cutting

1) Use <u>PAPER SCISSORS</u> to cut out <u>patterns</u> (take a look at p.54).

2) Use <u>DRESSMAKING SCISSORS</u> (also called <u>fabric shears</u>) to cut <u>fabric</u>. These have long, very sharp blades that cut through fabric more easily and neatly.

3) Use <u>EMBROIDERY SCISSORS</u> for more <u>delicate</u> jobs, e.g. snipping threads, or clipping curved seams to help press them. They have short, sharp blades.

4) Use <u>PINKING SHEARS</u> to cut fabric with a <u>zigzag edge</u> — this helps prevent fabric from fraying.

Use <u>CRAFT KNIVES</u> to cut <u>stencils</u> (e.g. if you want to <u>spray</u> a <u>design</u> onto fabric). You'll do a <u>neater</u> job than using scissors.

Use <u>SEAM RIPPERS</u> (quick unpicks) to <u>unpick</u> <u>seams</u>. Doing it by hand is slower, and scissors might accidentally cut the fabric.

Scissors and shears use shear forces to cut materials — see p.45.

For Sewing

1) You can use <u>pins</u> to <u>hold</u> the fabric together before stitching with a sewing machine.

2) They help keep your fingers <u>away from the needle</u> when you feed fabric through (be careful not to catch them with the machine though).

1) Use <u>NEEDLES</u> for <u>hand</u> stitches, e.g. <u>embroidery</u> stitches, attaching <u>beads</u> to fabric, or <u>tacking stitches</u>.

2) Use a needle that's the <u>right</u> <u>size</u> for the <u>thickness</u> of the <u>fabric</u> and the <u>thread</u> you're using.

There's more about pinning, tacking and using sewing machines on the next page.

For Measuring and Marking out

Use flexible <u>MEASURING</u> <u>TAPES</u> to accurately <u>follow</u> <u>curved</u> surfaces.

You can also use <u>TAILOR'S CHALK</u> and <u>PATTERN</u> <u>MASTERS</u> to help draw and mark out <u>patterns</u> (see p.149)

For Pressing

<u>DRY IRONS</u> use <u>heat and pressure</u> to press creases out of the fabric and flatten seams.

<u>STEAM IRONS</u> are more <u>effective</u> — they use <u>water and steam</u> as well as heat and pressure.

Irons can also be used to <u>apply designs</u> from <u>transfers</u> onto fabric (<u>heating</u> the transfer causes the design to <u>imprint</u> onto the fabric), or to <u>fix</u> designs done with <u>fabric crayons</u> and pens.

REVISION TASK

Standard components might pop up in the exam...

...so you should aim to remember the different types. Have a go at writing down everything you can remember from these two pages. If you're struggling, have another read over them and try again.

Joining and Shaping Fabrics

There are loads of different types of <u>machinery</u> that can be used to make <u>joining</u> and <u>shaping</u> fabrics <u>quicker</u> and <u>easier</u> to do. We'll cover these in a bit, but first off, we should talk about <u>methods</u> that are done <u>by hand</u>...

Pinning or Tacking is Used to Join Fabrics Together Temporarily

Fabrics can <u>slip</u> when you're trying to machine or hand sew them together, so it's best to <u>pin</u> or <u>tack</u> them in place first.

line you're going to sew along

Pinning

1) Pinning involves putting <u>pins</u> at <u>right angles</u> to the edge of the fabric.
2) If you're machine sewing, <u>remove</u> the pins as you come to them.

Tacking

1) Tacking (sometimes called <u>basting</u>) involves <u>hand sewing</u> long running stitches (about 1 cm). They should be in a <u>different colour</u> to the fabric, so you can easily see them.
2) Tacking holds the fabric together more <u>securely</u> than pinning (it's best to pin before tacking though).
3) Once you've tacked, remove all the <u>pins</u>, then you can see what your pieces will look like when joined. You can then <u>stitch</u> over your tacking.
4) Tacking stitches are <u>temporary</u> — they'll need to be <u>removed</u> once the stitching has been completed.

If your fabric will be <u>damaged</u> by pinholes or by tacks, e.g. PVC-coated or leather fabrics, you could use <u>paper clips</u> or <u>bulldog clips</u> to hold it in place instead.

Sewing Joins Fabrics Permanently

1) Once you've finished pinning or tacking your fabrics together, you're ready to start <u>sewing</u>.
2) <u>Hand-sewing</u> is great for <u>small, quick tasks</u> or those that need <u>precision</u>, such as embroidery or darning. For <u>bigger projects</u> it's quicker to use a <u>sewing machine</u>. They <u>speed up sewing</u>, and produce <u>neat</u>, <u>even</u> stitches for a <u>high-quality finish</u>.
3) Most sewing machines use two threads — one on a <u>bobbin</u> under the sewing plate, the other (the <u>top thread</u>) on the spool pin on top of the machine. The machine <u>interlocks</u> the two threads to make <u>stitches</u>.

1) <u>Before you start</u>, choose the <u>right needle</u> for your fabric and thread thickness and <u>fasten</u> it securely into the machine.
2) It's a <u>good idea</u> to do some lines of stitching on a <u>small sample</u> of your fabric first, so you can <u>check</u>:
   - That the thread <u>tension</u> is right. If the tension of the two threads is <u>balanced</u> then you'll get an <u>even stitch</u>, which <u>isn't too tight</u> or <u>too loose</u>.
   - That the stitch <u>type</u> and stitch <u>length</u> are correct.

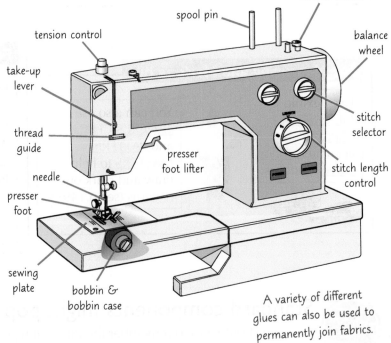

A variety of different glues can also be used to permanently join fabrics.

Joining and Shaping Fabrics

Overlockers can Sew Seams and Finish Edges at the Same Time

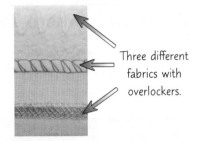

Three different fabrics with overlockers.

1) Overlockers are used to <u>finish edges</u> to <u>stop</u> them from <u>fraying</u>. They do this by <u>enclosing</u> the edge, or edges, in a thread casing.

2) An overlocker works by using several top threads, but no bobbin. It also has a <u>blade</u> to <u>trim</u> the fabric edge before it's enclosed.

3) They can be used just to finish seam edges, or to <u>sew, trim and neaten</u> the seam all in <u>one go</u>.

4) They can be used for <u>side seams</u> in <u>stretchy</u> clothes like T-shirts.

CAM can be Used to Cut and Sew Fabrics

CAM is covered in more detail on page 4 if you need to go over what it is.
It has lots of <u>different uses</u> within the <u>textiles industry</u>, from embroidery to cutting fabric...

Embroidery and Knitting Machines

1) <u>CAM embroidery machines</u> use <u>CAD</u> data to sew designs. They have many needles that change <u>automatically</u> as different coloured threads are needed.

2) <u>CAM knitting machines</u> use <u>CAD</u> data to control the <u>stitch pattern</u> and other design features. They can produce rolls of knitted fabric or even whole one-part garments.

3) <u>Computer control</u> makes these machines <u>very fast</u> and <u>accurate</u>, and the CAD <u>data</u> can be <u>quickly changed</u> to produce <u>new products</u>.

Cutting Machines

1) Before cutting, fabric is <u>automatically spread out</u> on the cutting table in <u>layers</u>.

2) A <u>CAM cutting machine</u> automatically cuts out the fabric pieces, following <u>CAD</u> lay plan instructions. (A lay plan is a plan that shows you how to position pattern pieces on fabric in the most efficient way.)

3) The machine cuts through <u>all the layers</u> at once, which makes the cutting process <u>really quick</u>.

4) It cuts the fabric <u>accurately</u> at <u>high speed</u> using vertical knives, high-pressure water jets or lasers (see p.5).

You can cut fabric by hand too using scissors and shears — see p.99.

Sewing Machines

1) <u>Industrial</u> sewing machines are very strong as they need to work at <u>high speeds</u>.

2) <u>CAM</u> sewing machines are used to carry out certain processes <u>automatically</u> — e.g. sewing buttonholes and attaching pockets.

Using CAM allows you to cut and sew very quickly...

Computers make light work of detailed design patterns — they can carry out complicated stitch patterns much more quickly than a human could. They are used in industry to produce garments on a large scale.

Joining and Shaping Fabrics

Now you've got to grips with the cutting and sewing basics, these pages cover the different types of <u>seams</u> and a few techniques you can use to <u>add shape</u> and <u>detail</u> to your work...

A **Seam** is where **Two Pieces** of Fabric are **Joined Together**

1) <u>Seams</u> are held together with <u>stitches</u>. They need to hold fabric <u>securely</u> and be <u>strong</u> enough to stand up to the <u>strains</u> put on the product.

2) There are different types you can use, depending on the <u>fabric</u> and what the product will be used for.

Plain (flat) seam — the simplest to do

1) <u>Plain</u> seams look <u>neat</u> on the outside — you can only see a thin joining line.

2) They are used for fabrics which <u>aren't</u> going to be under <u>too much strain</u>.

3) If your fabric is <u>stretchy</u>, use <u>stretch or zigzag</u> stitch to allow the seam to stretch.

A seam allowance is the extra material between the seam and the edge of the fabric — it's usually about 1.5 cm wide.

1) To make a <u>plain</u> seam, arrange two pieces of fabric so their outward-facing sides are touching. Then <u>pin</u> or <u>tack</u> to hold the fabric in place.

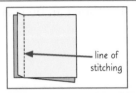

line of stitching

2) Stitch about <u>1.5 cm</u> in from the edge of the fabric (patterns have a <u>seam allowance</u> to allow extra fabric for this).

3) <u>Strengthen</u> the seam by <u>reversing</u> back over it for a few centimetres.

4) <u>Finish</u> the seam edges to <u>stop</u> them <u>fraying</u>. Then <u>open</u> out the seam and <u>iron</u> so it <u>lies flat</u>.

French seam — encloses the seam edges

1) French seams <u>enclose</u> the <u>raw edges</u> — they're used for <u>fine fabrics</u>, or fabrics which are <u>likely to fray</u>.

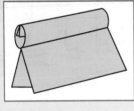

2) They're <u>strong</u>, but <u>not bulky</u>.

3) They're good for <u>baby clothes</u> — there are no rough edges to irritate the skin.

Flat-felled seam — strong and durable

1) Flat-felled seams have <u>two lines</u> of <u>stitching</u>. A plain seam is sewn and the <u>seam edges</u> are <u>enclosed</u> by wrapping one edge around the other. The whole thing is <u>ironed flat</u>, then a second line is stitched on top.

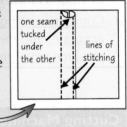

one seam tucked under the other

lines of stitching

2) They <u>stop</u> edges from <u>fraying</u> and their <u>strength</u> means they're used in hard-wearing clothes like <u>jeans</u>. You <u>wouldn't</u> use them with <u>delicate</u> fabrics because of the extra <u>bulk</u>.

Piping — adds definition to a seam

1) <u>Piping</u> can be used at <u>seams</u> to add <u>decoration</u> or to <u>strengthen</u> a product. It <u>stands out</u> from the seam, adding <u>definition</u>.

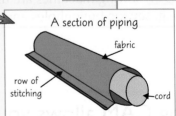

2) It is made from a <u>strip of material</u> that has <u>folded fabric</u>, or a <u>cord</u>, sewn inside.

3) It often has a bit of <u>spare fabric either side</u> so that it can be attached to a seam.

4) To add piping, <u>line up</u> your pieces of fabric on either side of the piping's seam allowance. Make sure that you lay your fabric <u>face down</u> onto the piping's seam allowance (so that you won't be able to see the seam once it's finished). Sew a <u>line of stitching over the fabric</u> — you can sew both pieces of fabric to the piping at once.

A section of piping

fabric

row of stitching

cord

5) Unfold your fabric and <u>iron the underside of the seam</u> — this will flatten the excess material and make it look neater.

Joining and Shaping Fabrics

Quilting uses Wadding

1) <u>Wadding</u> is any soft material that is used to <u>stuff</u> or <u>line</u> something.

2) <u>Quilting</u> uses <u>wadding</u> between <u>two layers</u> of fabric which are then <u>stitched</u> <u>together</u> in <u>straight lines</u> or in a <u>pattern</u>.

3) Quilting is often used to give added <u>warmth</u> to a product (e.g. an anorak or bed cover). The wadding <u>traps warm air</u> between the <u>layers</u> of fabric.

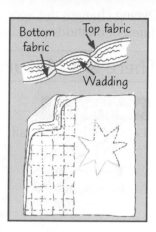

Bottom fabric · Top fabric · Wadding

<u>Advantages:</u>
1) Can create interesting <u>3D effects</u>.
2) Gives <u>warmth</u>.

<u>Disadvantages:</u>
1) Requires a <u>lot</u> of <u>material</u>.
2) <u>Time consuming</u>.

Gathering and Pleating are Shaping Techniques

You need to know some different techniques for <u>shaping</u> fabrics. Both <u>gathering</u> and <u>pleating</u> use <u>excess material</u> to create <u>detail</u>. They can also have <u>practical uses</u> — gathering can be used to give a <u>better fit</u>.

Gathers Pull in the Fabric Evenly along its Length

<u>Gathers</u> can be used at <u>waistbands</u> or <u>cuffs</u>.

1) <u>Knot</u> the threads at the start, or if machine sewing, leave a long thread that can be <u>wrapped around a pin</u>.

3) <u>Pull the threads</u> and ease the fabric along until it is drawn to the size you want.

2) Sew <u>two parallel rows</u> of stitches in the <u>seam allowance</u>. Use <u>small running stitches</u> when sewing by hand. If you're using a sewing machine, set the machine to its <u>longest stitch</u> and <u>loosest tension</u>.

4) If you want to <u>fix the gathering in place</u> (e.g. if you're making a skirt), knot the threads at the <u>end of the gathered fabric</u>.

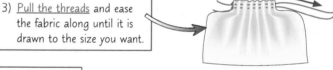

Pleats are Folds in Fabric

<u>Pleats</u> can be used to <u>shape</u> skirts.

1) Allow <u>three times</u> the finished width of fabric and <u>mark</u> the position of the pleats at <u>even widths</u> across the fabric.

2) <u>Fold</u> the pleats, <u>pin</u> them into position and stitch across the <u>top</u> of them to keep them in place.

3) <u>Pressing</u> gives the pleats <u>sharp creases</u>.

pleats

A pleated skirt

Pleats and gathers can make a garment look more interesting...

Lots of different techniques for joining fabrics on these pages — fascinating stuff. Combine that with your new knowledge of gathering and pleating and your prototype will be shaped and formed to perfection in no time.

Dyeing

Dyeing <u>changes the appearance</u> of a fabric — it can be used to <u>add colour</u> and <u>pattern</u>, and there's a few different methods you could use. Sometimes it even involves tea and onions — read on to find out how...

There are **Natural Dyes** and **Chemical Dyes**

Natural Dyes

1) Until the 1850s all dyes came from <u>natural sources</u>.

2) They're made from things like <u>onions</u>, <u>beetroot</u>, <u>tea</u>, <u>raspberries</u> or <u>flowers</u> etc.

Fibres and fabrics can be dyed at <u>different stages</u> of the manufacturing process — individual <u>fibres</u>, <u>yarns</u> or whole pieces of <u>fabric</u> can all be dyed. The methods here are for <u>dyeing fabrics</u>.

Chemical Dyes

These were invented in the <u>1850s</u>.

<u>Advantages</u>
- Colours are <u>brighter</u>.
- <u>Easier</u> and <u>cheaper</u> to make.
- Exactly the same colour can be achieved <u>repeatedly</u>.

<u>Disadvantage</u>
Some are <u>toxic</u> — they can be <u>harmful</u> to <u>people</u> and the <u>environment</u>.

Some Fabrics are **Better** for Dyeing than Others

1) <u>Natural</u> fibres (like cotton, wool and silk) are the <u>best</u> for dyeing as they're very <u>absorbent</u>.

2) The <u>colour of the fabric</u> you begin with makes a difference to the final colour — for example if you dye <u>white fabric red</u>, you get <u>red fabric</u> — if you dye <u>yellow fabric red</u>, you get <u>orange</u>.

3) Fabrics that have an <u>uneven colour</u> need to be <u>bleached</u> before dyeing to ensure an even colour.

4) For some fabrics and dyes you need to use a chemical called a <u>mordant</u> (e.g. salt) to <u>fix</u> the colour to the fabric. This makes the fabric <u>colour-fast</u> — the dye won't come out in the wash.

Batch Dyeing is a **Commercial** Dyeing Technique

1) Commercial dyeing involves dyeing a <u>huge amount</u> of fabric at a time. The fabric is dyed a <u>uniform colour</u>. It can be done <u>continuously</u> — this is where <u>very long lengths</u> of fabric are dyed the <u>same</u> colour in a continuous process. Or it can be done in <u>batches</u>...

2) In <u>batch dyeing</u>, a <u>batch</u> of fabric (a specific amount) is dyed with <u>one colour</u>, then another batch with a different colour, and so on. Here's one way it's done:

1) The fabrics are initially produced and <u>stored without dyeing</u>.

2) When a <u>batch</u> of coloured fabric is needed, the <u>required amount</u> is dyed.

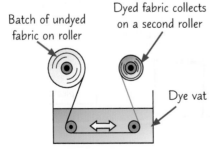

Batch of undyed fabric on roller

Dyed fabric collects on a second roller

Dye vat

Fabric passes back and forth between rollers

3) The fabric is <u>mounted on two rollers</u> and passed <u>back and forth</u> through the dye until all the dye is <u>used up</u>.

4) The dyed material is put in a separate machine to <u>fix</u> the dye (to stop the colour from running and to <u>wash off</u> excess dye).

During the fixing process, dye is often oxidised (it gains oxygen). This can be done by exposing the dye to oxygen in the air or using a specific chemical. Oxidation causes the dye to become insoluble, so it won't dissolve in water and come out in the wash.

3) Batch dyeing is useful in industry, as it allows textiles <u>manufacturers</u> to <u>respond quickly</u> to orders for a <u>specific colour</u> of fabric.

Dyeing

Hand Dyeing — Use a Resist to make Patterns

One of the <u>advantages</u> of <u>hand</u> dyeing is that you can add <u>designs</u> to the fabric using a 'resist'. A 'resist' is something which <u>prevents the dye</u> from reaching the <u>fabric</u> — it's applied in a pattern before dyeing.

Tie-Dye is Easy to do

Different ties give different patterns

1) Fabric is <u>tied</u> with <u>string</u> or <u>rubber bands</u> to create a <u>resist</u>.

2) The fabric is then <u>immersed</u> in dye.

3) The dye <u>doesn't get to</u> the tied areas.

4) Once the dye has <u>dried</u> and the fabric is <u>untied</u>, the pattern is <u>revealed</u>.

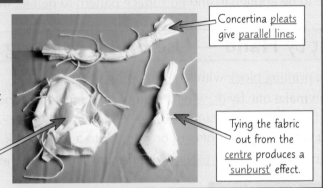

Concertina <u>pleats</u> give <u>parallel lines</u>.

<u>Small circles</u> are formed when <u>pebbles</u> or <u>buttons</u> are tied in.

Tying the fabric out from the <u>centre</u> produces a '<u>sunburst</u>' effect.

This vest has been tie-dyed.

Advantages
1) Every piece will be <u>unique</u>.
2) The <u>equipment</u> is <u>cheap</u> and <u>readily available</u>.

Disadvantages
1) The outcome is <u>unpredictable</u>.
2) You <u>can't repeat</u> a pattern exactly.
3) You can't create a <u>detailed pattern</u>.
4) It's <u>time-consuming</u> for <u>large</u> areas.

Batik — Good for more Detailed Designs

In <u>batik</u> the resist used is usually <u>hot wax</u>.

1) The fabric is stretched across a <u>frame</u>. Then the <u>hot wax</u> is applied with a <u>brush</u> or a <u>tjanting</u> (pointed tool for dripping the wax) to create a pattern.

2) Once the wax is dry, the dye is <u>painted</u> on, or the fabric can be immersed in a <u>dye bath</u>.

3) The wax is <u>ironed</u> off to reveal the pattern.

Advantages
1) It's a more <u>precise</u> way of adding <u>patterns</u> to fabrics than tie-dye.
2) <u>Patterns</u> can be <u>more detailed</u>.
3) Every product will be <u>unique</u>.

Disadvantages
1) It's <u>time-consuming</u> — each part of the design has to be painted separately.
2) You need to be <u>careful</u> when working with <u>hot wax</u>, and it's easily dripped in the <u>wrong</u> place.
3) It can be <u>tricky</u> to iron all the wax out.

Batik works <u>best</u> on <u>natural</u> fabrics — they <u>absorb</u> hot wax and cold dye more easily than synthetics do.

Tie-dye and batik are good ways to make one-off items...

Make sure you remember the different dyeing methods that we've covered on these pages — there's quite a lot of detail to learn about each one. And after all, it could be easy marks in the exam.

Section Six — Textiles

Printing

Just like with dyeing, there are different methods of printing that can be done <u>by hand</u> or <u>machine</u>. You'll need to understand <u>how they work</u>, and why you might use each type. Lots to learn. Here we go...

Printing is Used to Apply a Design to Fabric

1) Printing is the process of applying <u>ink</u>, <u>dye</u> or <u>paint</u> to fabric in <u>defined patterns</u>.
2) Printing can be done in <u>small quantities</u> by <u>hand</u>, or in <u>large quantities commercially</u>.
3) Materials with a <u>tight weave</u> are best for printing on because they have a <u>smooth surface</u> for the dye to be applied to and <u>no surface pattern</u> to <u>detract</u> from the design.

A printed dress

To Print by Hand — Use Block Printing...

You need a <u>printing block</u> with a <u>raised design</u>.

1) You can make one by <u>drawing</u> a pattern on a piece of wood and then <u>cutting</u> the <u>background away</u>, leaving the design <u>raised</u>.
2) Or you can <u>stick</u> pieces of <u>card</u> or <u>string</u> onto a solid <u>block</u>.
3) You can also <u>buy rubber blocks</u> with designs already on them.

The Printing Process...

1) <u>Printing ink</u> is applied to the <u>raised surface</u> of the block.	2) The block is then <u>pressed down</u> onto the fabric...	3) ...leaving a <u>reversed</u> image on the fabric.

Advantages

1) You can print with <u>different colours</u> using the <u>same block</u>.
2) You can use <u>several blocks</u> to build up a more <u>complicated</u> design.
3) You can easily <u>repeat</u> designs.
4) You can use the same block <u>many</u> times before it <u>wears out</u>.

Disadvantages

1) <u>Making</u> the blocks takes a <u>long time</u>.
2) It's not good for <u>fine detail</u>.

In industry, <u>engraved rollers</u> are used instead of blocks. The rollers are inked as the fabric is <u>continuously</u> run under them. The rollers are <u>expensive</u> to make, so it's only cost-effective for large amounts of fabric.

...or Flat-Bed Screen Printing

1) The <u>screen</u> is a <u>frame</u> with <u>fine mesh</u> covering it.
2) A <u>stencil</u> is cut from card or acetate by hand or using CAD/CAM and put <u>beneath</u> the screen (on top of the fabric). Alternatively, parts of the screen itself can be <u>blocked off</u> to create the design.
3) Printing <u>ink</u> is poured onto the screen.
4) A <u>squeegee</u> is pressed down and drawn across the screen, forcing the ink <u>through</u> the mesh and the holes in the stencil.
5) The screen is <u>lifted up</u> and the design is left on the fabric.

Advantages

1) Can produce <u>intricate patterns</u> that can be <u>repeated accurately</u>.
2) Good for printing <u>large areas</u> of colour.
3) Easy to use <u>many colours</u> — by using several screens.
4) Printing is <u>quick</u> to do (once you have the screens and stencils).

Screen printing can also be done by a machine — see the next page.

<u>Flat-bed Screen</u> printing by <u>hand</u>...

Squeegee

Ink

Mesh

Disadvantages

1) <u>Making</u> the <u>screens</u> and stencils takes a long <u>time</u>.
2) Each colour has to be applied <u>separately</u>.

Printing

Screen Printing is done Industrially with Machines

1) Industrial screen printing uses <u>machines</u> to repeat the <u>same pattern</u> all the way along a <u>long length</u> of fabric.

2) There are <u>two</u> methods of industrial screen printing. Like screen printing by hand, they both use some sort of <u>screen</u> and a <u>squeegee</u>.

The screens can be made using CAD and CAM.

Flat-bed screen printing

1) This is similar to flat-bed screen printing <u>by hand</u>. <u>Several</u> screens are used — one for <u>each</u> colour.

2) The fabric passes <u>under</u> the screens on a <u>conveyor belt</u> and the colours are applied one after the other.

Between each ink application, the screens are <u>lifted</u> and the fabric <u>moves along</u> the belt to the next screen position.

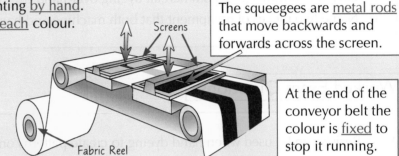

Screens

Fabric Reel

The squeegees are <u>metal rods</u> that move backwards and forwards across the screen.

At the end of the conveyor belt the colour is <u>fixed</u> to stop it running.

Rotary screen printing

1) Rotary screen printing is similar to flat-bed screen printing but the <u>screens</u> are on <u>cylinders</u>.

2) This is a really <u>fast</u> and <u>widely used</u> industrial method of printing.

3) <u>300 m</u> of fabric can be printed every <u>minute</u>.

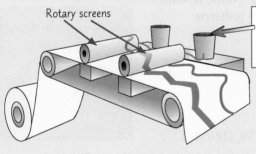

Rotary screens

A <u>different colour dye</u> is pumped into each of the rotary screens.

The dye is squeezed through holes in the <u>rotary screen</u> by a <u>cylindrical squeegee</u>.

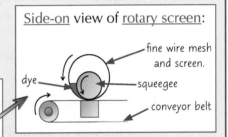

<u>Side-on</u> view of <u>rotary screen</u>:

fine wire mesh and screen.

dye

squeegee

conveyor belt

Advantages

1) Just as with <u>hand</u> screen printing, <u>intricate patterns</u> using <u>many colours</u> can be <u>repeated</u> accurately.

2) The <u>machinery</u> makes the printing process <u>much quicker</u> and <u>large amounts</u> of fabric can be printed at once. Rotary printing is the <u>quickest</u> way — as fabric is passed along the conveyor belt <u>continuously</u>.

3) The <u>screen making</u> process can be <u>computer-controlled</u>, which makes it much <u>quicker</u>.

Machines can also print on individual garments, e.g. T-shirts.

Disadvantages

1) Setting up commercial <u>machinery</u> is <u>expensive</u>.

2) <u>Screen making</u> takes a <u>long time</u>, unless it's done using CAD/CAM.

EXAM TIP

Loads of industrial printers use screen printing...

In the exam you might get asked to compare different methods or techniques (e.g. block printing and screen printing). When making comparisons you need to write about the similarities and differences between different methods — to do this you need to really know your stuff.

Warm-Up and Worked Exam Questions

Section Six is nearly all sewn up — just a few question pages to go. You probably know the drill by now.
Try the warm-up questions, and read through the worked exam questions before tackling the exam questions.

Warm-Up Questions

1) Name the process that is used to construct a fabric with two or more layers.
2) Describe briefly how Velcro® works as a fastening.
3) What is the purpose of a seam?
4) Give one advantage of commercial dyeing over hand dyeing.
5) Name two pieces of equipment that both machine and hand screen printing methods use.

Worked Exam Questions

1   Resists can be used when hand dyeing to create patterns on the fabric.

a)   Describe how a resist is used in the process of hand dyeing.

A resist is applied in a pattern before dyeing and prevents the dye from

reaching the fabric.

[1 mark]

b) i)   **Figure 1** shows a design created by hand dyeing. Name a hand
dyeing method that could be used to create these patterns.

batik

[1 mark]

ii)   Briefly describe the main steps in this method.

Hot wax is applied with a brush or tjanting to create a

pattern. When the wax is dry, the dye is painted on.

Then the wax is ironed off to reveal the pattern.   Alternatively, the fabric could be
immersed in a dye bath to apply the dye.

Figure 1

[3 marks]

2   A sports T-shirt is to be made from 100% polyester.

Give **two** properties of polyester and explain why each one makes it appropriate for sports use.

1. It is strong, so the T-shirt won't easily be damaged during sports.

It's also resistant to abrasion for a similar reason.

2. It dries very quickly, so is suitable for outdoor use.

This also means that sweat dries quickly, so it's a suitable material for sports clothing.

[4 marks]

Exam Questions

1 A seamstress is buying materials to make a shirt. Three different fabrics that could be used for the shirt are shown in the table below.

Fabric	Width (cm)	Cost per metre (£)	Length needed (m)	Total cost of fabric (£)
A	132	5.50	1.70	
B	150	7.40	1.55	
C	150	6.20	1.55	9.61

a) Complete the table by calculating the total cost for fabrics A and B.

....

....

[1 mark]

b) i) The seamstress decides to use **fabric C**. However, a change is made to the design of the shirt, which means the length of material needed increases to 1.90 metres. Calculate the new total cost of fabric required to make the revised design.

....

[1 mark]

ii) Calculate how much more of fabric C is required for the revised design compared to the original design. Your answer should be given in m^2 and to two decimal points.

....

....

....

[3 marks]

2 Fabrics can be stitched together by hand or using a sewing machine.

a) i) When are sewing machines better suited to stitching fabrics than hand-sewing? Give a reason for your answer.

....

....

[2 marks]

ii) Explain when hand-sewing is more suitable for stitching fabrics than a sewing machine.

....

....

[2 marks]

b) A sewing machine is being prepared for use. The machine is used to sew a line of stitches in a spare piece of fabric. Suggest why this is done.

....

....

[1 mark]

Exam Questions

3   Acrylic fibres are often used in furnishings such as carpets.

Give **two** properties of acrylic fibres that make them suitable for use in carpets.

1. ....

2. ....

[2 marks]

4   A batch of 40 T-shirts are to have a simple repeating
star design like the one shown in **Figure 2**.
The stars are to be printed in different colours, and
they are to be printed using a block printing method.

a)   Using sketches and/or notes, explain how block
printing can be used to make this T-shirt.

Figure 2

[4 marks]

b)   Suggest **two** advantages of using block printing for this batch of T-shirts.

1. ....

2. ....

[2 marks]

Revision Questions for Section Six

Hurrah, you've reached the end of <u>Textiles</u> — time to see if you know your gathers from your pleats.
- Try these questions and <u>tick off each one</u> when you <u>get it right</u>.
- When you've done <u>all the questions</u> for a topic and are <u>completely happy</u> with it, tick off the topic.

Fabrics and Their Properties (p.96-97) ☐

1) Name two properties of polyamide fabric which make it suitable for use in sportswear. ☑
2) State a property of cotton that makes it suitable for use in cushions.
   Give a reason for your answer. ☑
3) Give an example application of a flame retardant treatment. ☑
4) Name a treatment that could be used on a dining room carpet. Give a reason for your answer. ☑

Standard Components and Tools (p.98-99) ☑

5) A fabric is sold from a roll that is 90 cm wide. Calculate the surface area of a 3 m length. ☑
6) Give one advantage and one disadvantage of each of the following components:
   a) zips   b) Velcro® ☑
7) Which type of scissors cut a fabric in a way that helps prevent it from fraying? Explain how. ☑

Joining and Shaping Fabrics (p.100-103) ☑

8) What is a tacking stitch used for? ☑
9) Describe how a sewing machine forms stitches. ☑
10) Why is an overlocker used to finish edges? ☑
11) a) Give an example of one process that a CAM sewing machine could carry out automatically.
    b) Suggest another piece of textiles equipment that uses CAM. ☑
12) State which types of seam would be most suitable in the following items of clothing and explain why.
    a) baby clothes   b) jeans ☑
13) John wants to add decoration to his sofa cushions. Suggest a technique he could use to do this. ☑
14) Give an example of where a gather can be used in clothing. ☑
15) Give one advantage and one disadvantage of quilting. ☑

Dyeing (p.104-105) ☑

16) Why might a chemical dye be chosen over a natural one? Give two reasons. ☑
17) What are mordants used for? Give an example of a mordant. ☑
18) Name a process that allows large quantities of fabric to be dyed at once. ☑
19) What type of resist might you use in batik? ☑

Printing (p.106-107) ☐

20) Why are materials with a tight weave better for printing on? ☐
21) Sophie wants to block-print some letters onto her fabric.
    Describe one way in which she could make her own printing block. ☑
22) What is the squeegee used for in flat-bed screen printing? ☐
23) Give one advantage and one disadvantage of using industrial screen-printing techniques. ☑

Properties of Components in Systems

Electronic and mechanical components are made from materials that have specific properties.

Systems are Made from Materials with Suitable Properties

1) Mechanical systems are made up of lots of individual physical parts. In much the same way, electronic systems contain many electrical components. Each one has a particular function in the system.

2) Different components in a system are commonly made from different materials. This is because we often require materials with a certain set of properties to perform a specific function.

Motor Vehicles

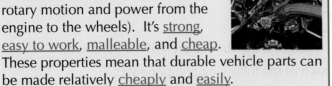

1) Steel is often used for vehicle parts, e.g. the brake discs, engine, wheels, and drive shaft (a rod that transmits rotary motion and power from the engine to the wheels). It's strong, easy to work, malleable, and cheap. These properties mean that durable vehicle parts can be made relatively cheaply and easily.

2) Car bodies are also made from steel because it's tough. In an accident it's designed to crumple, absorbing most of the impact — this helps protect the passengers.

3) Aluminium is a more corrosion-resistant alternative to steel for car body parts. It's also less dense, so aluminium parts are lighter than steel — this improves performance (e.g. acceleration and fuel economy). It is more expensive and harder to work though, so it's more commonly used in expensive vehicles.

4) Strips of electrically conductive material on the inside of a car's rear window can be used to defog it. When an electric current is passed through the strips, they heat to around 30 °C and clear the glass.

5) Engines get hot when in use so engine electronics are made from materials that can cope with high temperatures. These electronics have to be able to work in freezing conditions too.

Domestic Appliances

1) Washing machines use a drive belt — a band that transmits rotary motion from the motor to the spinning drum. They're often made from rubber or a polyurethane plastic, which is:

   - Strong — able to resist pulling forces without breaking.
   - Flexible so the belt can bend.

   The spinning drum, and any pumps and valves, must be resistant to corrosion as they're exposed to water during use — stainless steel (see p.22) or plastic is often used.

2) A heating element is a wire that gets hot when a current passes through — this is caused by the electrical resistance of the element. They can be found in appliances such as toasters, kettles and electric ovens.

3) Heating elements are often made of a nickel-chromium alloy. It has a high melting point (so it can get hot without melting) and is corrosion-resistant (so it can cope with being exposed to air and water).

Properties of Materials can be Modified for Specific Purposes

Photosensitive Material in Printed Circuit Boards

1) Photosensitive materials respond to being exposed to light. For example:

   - The photographic film in a camera goes through a chemical change that is dependent on the amount of light present. This can be developed into a photo.
   - A light-dependent resistor contains a material with an electrical resistance that varies with light levels (see p.31).

2) When making printed circuit boards (PCBs), a process called photo-etching is used to transfer the design of the circuit onto the board.

3) The circuit board initially has a photosensitive layer on top of a layer of copper. When this layer is exposed to UV light it is modified — this allows copper to be removed from specific bits of the board (see p.118).

Properties of Components in Systems

Anodised Aluminium

1) Some metals (such as aluminium) naturally <u>react</u> with <u>oxygen</u> in the <u>air</u> to form a <u>layer</u> of <u>metal oxide</u> on <u>exposed surfaces</u> — this is known as <u>corrosion</u> or <u>oxidation</u>.

2) When aluminium oxidises, it forms <u>aluminium oxide</u>. Unlike rust, this is a <u>hard</u> material that <u>sticks</u> strongly to the fresh aluminium beneath, <u>protecting</u> it from <u>further corrosion</u>.

3) <u>Anodisation</u> is a process that <u>thickens</u> the <u>protective</u> oxide layer that is already present using an <u>electric current</u>. In <u>aluminium</u>, this makes the surface <u>harder</u> (and even more <u>resistant to corrosion</u>).

4) Anodised aluminium is used to make loads of stuff including <u>aircraft parts</u>, <u>pans</u> and <u>window frames</u>.

Iron and steel oxidise in the presence of air and water. This causes rust to form on the surface of the metal.

Anodised aluminium pans have a hardened surface that's difficult to scratch. This makes them durable and means you shouldn't get aluminium in your food.

Current and Voltage Ratings Prevent Damage to Components

1) Most <u>electronic components</u> have a <u>current</u> and <u>voltage rating</u> — these are the <u>current</u> and <u>voltage values</u> that the component is <u>designed</u> to <u>work at</u>.

2) They're <u>set</u> by the <u>manufacturer</u> and should be <u>followed</u> to <u>avoid damage</u> being done to the component.

Remember, the voltage from a power supply pushes an electric current around a circuit.

mA stands for milliamp — 1 mA is the same as one thousandth of an amp (0.001 A).

For example, a <u>light-emitting diode</u> (LED) will typically have a current rating of around <u>20 mA</u> (0.02 A) and a voltage rating of around <u>2 V</u>. If the current and voltage are a lot <u>smaller</u> than this (e.g. 2 mA or 1 V), the LED may be <u>dim</u> or <u>not light up at all</u>. However, if we <u>go much over</u> the ratings (e.g. 1 A and 10 V), the LED will <u>heat up</u> and become <u>damaged</u>.

3) Components should be <u>bought</u> to <u>match</u> a chosen <u>power supply</u>. For example, <u>mains</u> power supplies a <u>high voltage</u> (<u>230 V</u>), so is often used for <u>power-hungry</u> components. <u>Batteries</u> commonly have a <u>lower voltage</u> than mains power and are usually used in <u>smaller</u> or <u>portable</u> products.

Battery Type	Voltage Supplied	Applications
AA or AAA	1.5 V	Alarm clocks, remote controls, torches
PP3 (9-volt battery)	9 V	Smoke detectors, walkie talkies, radios, microphones
Button cells	1.5 - 3 V	Watches

4) <u>Resistors</u> are often used in circuits to <u>modify</u> the <u>current</u> from a power supply (see p.30) — this makes sure components are operating at their <u>current rating</u> and will not be <u>damaged</u>.

It's all about picking a component with the right properties for the job...

...and failing that, you can always modify the properties of a material to make it more suitable. Make sure you learn the different ways that the properties of materials can be modified — it could be easy marks in the exam.

Standard Components in Systems

There are loads of <u>standard components</u> in <u>electronic</u> and <u>mechanical systems</u>. We'll cover just a <u>few</u> here...

A **Common** Set of **Resistors** is the **E12 Series**

1) <u>Resistors reduce the current</u> flowing round a circuit. E12 resistors are <u>fixed resistors</u> — this means they have a <u>constant resistance</u>, unlike LDRs and thermistors (see p.31).

Resistance is how much a component <u>limits</u> the <u>current</u> flowing through it.

2) Manufacturers don't make resistors for <u>every</u> single possible <u>resistance</u> that a customer could want — instead, they make different <u>series</u> of resistors, which offer different <u>sets</u> of resistance values.

3) In the <u>E12 series</u>, there are <u>12</u> different values of resistance — these are 1.0, 1.2, 1.5, 1.8, 2.2, 2.7, 3.3, 3.9, 4.7, 5.6, 6.8 and 8.2 ohms (Ω). These can also be <u>multiplied</u> by a <u>power of 10</u>, e.g. 1.2 Ω × 103 gives a 1200 Ω resistor.

4) E12 resistors have a <u>tolerance</u> of ±<u>10%</u> — this means that the <u>actual resistance</u> is <u>within 10%</u> of the <u>value</u> stated on the resistor.

± stands for plus or minus. E.g. 100 ± 10% means the value could be anywhere from 10% less than 100 to 10% more than 100.

EXAMPLE: **A 100 Ω E12 resistor has an actual resistance of 100 Ω ± 10%.**

This means that the actual resistance of the resistor is in the range of 10% below 100 to 10% above 100.

10% of 100 = 10. So the actual resistance is between 90 Ω (100 − 10) and 110 Ω (100 + 10).

Colour of band	Number
Black	0
Brown	1
Red	2
Orange	3
Yellow	4
Green	5
Blue	6
Violet	7
Grey	8
White	9

If your circuit needs a 105 Ω resistor, you would normally choose a 100 Ω resistor as 105 Ω falls <u>within</u> the ±10% tolerance. However, you may want to play it <u>safe</u> and choose a 120 Ω resistor (which has resistance between 108 and 132 Ω) — this would <u>make certain</u> that the <u>current isn't too large</u>, as the <u>resistance</u> will <u>definitely be higher</u> than what you <u>need</u> (it'd be at least 108 Ω).

5) <u>Coloured bands</u> are used to <u>identify</u> resistors. These are read from <u>left to right</u> (when the tolerance band is on the far right). The <u>colour</u> of each band <u>represents</u> a <u>number</u>.

6) There are <u>4</u> coloured bands on E12 resistors:

- <u>Bands 1 and 2</u> represent the <u>first two digits</u> of the value of <u>resistance</u>.

- <u>Band 3</u> represents the <u>number of zeros</u> that <u>come after</u> the <u>first two digits</u> in the resistance value.

- <u>Band 4</u> represents the <u>tolerance</u> — it's always <u>silver</u> (representing <u>10%</u>) for <u>E12 resistors</u>.

EXAMPLE:  **Identify a resistor with a yellow, violet, red and silver stripe.**

Yellow = 4, violet = 7, red = 00 and silver = 10% tolerance.
The resistor = 4700 Ω with a 10% tolerance.

Dual in Line IC Packages Have **Two Rows** of **Pins**

1) <u>Integrated circuits</u> (ICs) are <u>tiny</u> circuits that are <u>contained</u> within a <u>single component</u> (see p.32).

2) A <u>Dual in Line package</u> (<u>DIL</u>) is <u>one</u> of the ways that <u>ICs</u> are <u>packaged</u>:

- The circuit itself is placed inside a <u>rectangular protective casing</u> made of <u>plastic</u> or <u>ceramic</u>.
- <u>Metal legs</u> (<u>pins</u>) stick out of the casing in <u>two parallel rows</u>. The <u>number of pins varies</u> depending on the type of IC — commonly <u>between 8 and 40 pins</u>.

3) The pins <u>connect</u> the <u>IC</u> to the <u>rest of the circuit</u> — they're <u>pushed</u> through <u>drilled holes</u> in the PCB and are <u>soldered</u> on the <u>other side</u>. Alternatively, they can be <u>plugged into</u> an <u>IC socket</u> (a component that an IC slots into) so you don't have to <u>solder</u> the IC <u>directly</u> to the board — this means they can be <u>easily replaced</u>.

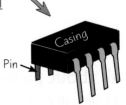

Casing
Pin

4) Each pin has a <u>number</u> — a <u>dot</u> or <u>notch</u> on the casing shows where the <u>1st pin</u> is, and they're numbered <u>anticlockwise</u> from that pin. This is useful as <u>most</u> pins have a <u>certain function</u>, e.g. some <u>only</u> connect to inputs.

Standard Components in Systems

PICs are a Specific Type of Microcontroller

1) Microcontrollers are programmable ICs that can do the job of multiple process blocks (see p.32) — a programmable intelligent computer (PIC) is a common type of microcontroller.

2) PICs are often sold as DILs (Dual in Line packages) and have a voltage rating of between 3 and 5.5 V.

3) Some PICs are one-time programmable (OTP) microcontrollers — they can only be programmed once.

4) However, most PICs have flash memory — they can be reprogrammed up to 100 000 times, so can be used again and again.

So when choosing a PIC, keep in mind that:

- Different PICs have different amounts of memory, types of memory and numbers of pins.
- PICs with more memory and more pins cost more to buy.

PICs with more pins can have more inputs and outputs.

Mechanical Components Come in Loads of Shapes and Sizes

Chain   Sprocket

Chains

1) Roller chains like the one on a bike are made up of metal links that are pinned together to form a loop. Chains transfer rotary motion from one sprocket to another.

2) Chains are sold by width and pitch (distance between the centres of the pins).

3) Chains only come in standard lengths — you may have to add or remove some links to make it the length you need.

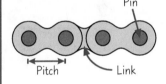

Pin

Pitch   Link

Gears

1) Gears come in a variety of widths.

2) They can also have different numbers of teeth — this is important in a gear train to change the speed of rotation (see p.35).

3) Different types of gear will have different tooth shapes that interlock in different ways — for example, bevel gears have teeth angled at 45°.

Slower rotation   Faster rotation

Springs

1) There are 2 major types of springs that you can buy — compression springs (which resist compression) and extension springs (which, that's right, resist extension).

2) Springs can be found with different lengths, widths, and numbers of coils.

3) This can affect the spring rate of the spring, i.e. how easy it is to stretch or compress.

Extension

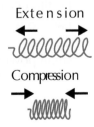

Compression

EXAM TIP

ICs are used in all kinds of tech — phones, cars, computers...

ICs, DILs, PICs, OTPs — there are some tricky abbreviations to get your head around in this topic. Once you've learnt them though, they can be really useful. For example, in the exam, it'll save you writing out the long-winded version each time. No more aching hands for you...

Cutting, Drilling and Soldering

Cutting and drilling processes are the bread and butter of most manufacturers — they're used all the time when working with metals, wood and plastics (see Section Five). Here are some examples that focus on systems...

Laser Cutters Cut Out System Components

1) Laser cutters use a laser to cut through materials.

2) The machine makes cuts by following a design that's loaded into it. This instructs it where to cut the material.

3) Laser cutters use designs made by CAD — they are CAM machines.

> Computer-aided design (CAD) involves designing products using specialist computer software. Computer-aided manufacture (CAM) produces products from a design that's made using CAD (see p.4).

> Like all CAM machines, laser cutters use computer numeric control (CNC). This involves using the precise coordinates set out in the design to control exactly where the laser cuts the material.

4) Laser cutters can only cut in 2D, so they have to be used on sheets of material — these can be sheets of plastic, wood, cardboard, fabric or some metals. Thicker sheets may need to be cut with more powerful industrial lasers.

5) Laser cutting has a huge range of uses — it can be used to cut out a variety of mechanical components (such as gears), and PCBs too.

6) It has high precision and accuracy — it's able to follow complex patterns, even on a small scale. This is helped by the tiny width of the laser beam.

7) However, a laser cutter requires a lot of power to run, so it can be expensive.

PCB Drills Cut Holes in PCBs

1) Printed circuit boards need to be drilled so that the pins or wires of each component can be pushed through the holes and soldered in place (see next page).

2) Printed circuit boards are often very small, so they need tiny holes. As a result, very thin drill bits are used — they can be as little as 0.2 mm wide.

3) Drilling PCBs can be a fiddly business — holes have to be drilled perfectly straight and in exactly the right spot. We therefore use a drill press (a small power drill that has a stand) rather than a hand-held drill to make it easier to drill the holes.

4) In industrial PCB production, CAM drills are often used, as they make the process even quicker and easier to do.

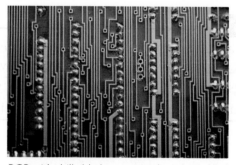

PCB with drilled holes — some have been filled with solder to attach components.

Laser cutters are ideal for certain jobs, but can be expensive in the long run...

There's some interesting stuff on this page, so hopefully it won't be a chore to learn. Make sure you know how computer-controlled laser cutting works, and that laser cutters and some PCB drills are CAM machines.

Cutting, Drilling and Soldering

Soldering is used to join metals together — its most common application is in mounting components onto PCBs, so we'll be focusing on this here. Other applications include joining plumbing pipes and sealing food cans.

Soldering Provides an **Electrical Connection** in **Circuits**

1) Soldering uses solder to attach components to a circuit — it's used to connect a component's metal pins to either wires or the copper tracks on a PCB, and hold the components in place.

Some components don't have pins or wires. They are directly mounted to the surface of the PCB.

2) Solder is usually a tin-based alloy, and can be applied manually by melting it with a hot soldering iron.

3) It's electrically conductive, so it's used as an electrical connection between the components and the rest of the circuit.

Soldering iron

Soldering is also covered on p.90.

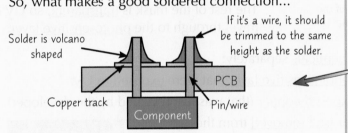

So, what makes a good soldered connection...

Solder is volcano shaped

If it's a wire, it should be trimmed to the same height as the solder.

Copper track

PCB

Component

Pin/wire

Badly soldered joints can have balls of solder rather than volcano shapes — these make poor electrical connections. Bad soldering can also involve using too much solder — this connects copper tracks that shouldn't be, which alters the circuit.

4) In industry, it would be too slow to mass produce PCB circuits using manual soldering. So automated soldering methods are used instead:

Flow Soldering

* This is when components are placed onto the PCB, which is then passed over a pan of molten solder. Sometimes components are glued into place before being passed over the molten solder.

* A pump produces an upwelling, or 'wave' of solder at the right height to just touch the base of the board and solder each component in place.

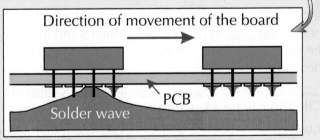

Direction of movement of the board

PCB

Solder wave

Flow soldering can also be called "wave soldering".

* The whole process is quick and cheap, but a lot of thought has to go into component layout to make it work.

* It's generally only used for components with pins that poke through the PCB, e.g. DILs.

Flow soldering lets you produce soldered PCBs really quickly...

Soldering serves two really important purposes in PCBs — it's used to hold components in place on the board and provides an electrical connection between the components and the circuit itself.

PCB Production and Surface Treatments

It's time to <u>finish</u> PCBs off once and for all. It's not as bad as it sounds... just <u>etching</u>, <u>assembly</u> and <u>lacquering</u>.

Photo-Etching Removes any Copper we Don't Want as Tracks

<u>Photo-etching</u> converts a <u>blank PCB</u> (with a layer of copper over the <u>whole</u> board) into one with <u>custom copper tracks</u> that is <u>ready</u> for <u>drilling</u> and <u>soldering</u>. The process involves <u>UV light</u> and a <u>bucketful of chemicals</u> — it <u>removes all of the copper</u> from the board <u>except</u> for the <u>bits</u> that are needed to make the <u>tracks</u>.

1) The <u>layout</u> of the copper tracks is <u>designed</u> using <u>CAD</u> software — this is <u>printed</u> onto a sheet of <u>transparent material</u> (e.g. a plastic such as <u>acetate</u>) to form a "<u>PCB mask</u>".

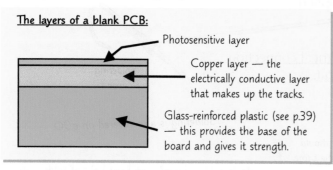

The layers of a blank PCB:

Photosensitive layer

Copper layer — the electrically conductive layer that makes up the tracks.

Glass-reinforced plastic (see p.39) — this provides the base of the board and gives it strength.

2) The mask is placed onto the blank PCB on top of the <u>photosensitive layer</u> (see p.112).

3) The <u>board</u> is then <u>exposed</u> to UV for <u>3-10 minutes</u>. The <u>mask</u> is <u>opaque</u> in the <u>areas</u> of the board that will become <u>copper tracks</u>, so it <u>shields</u> these parts from the UV. The rest of the mask is <u>transparent</u>, so UV can <u>travel</u> through to the <u>photosensitive layer</u>.

4) The board is then exposed to two different <u>chemicals</u> separately:

- <u>Developer</u> — this <u>dissolves</u> parts of the <u>photosensitive layer</u> that were <u>exposed to UV</u>.
- <u>Etching chemical</u> — this <u>dissolves</u> any <u>exposed copper</u> (that was just revealed by the developer).

5) Copper <u>still covered</u> by a photosensitive layer <u>isn't removed</u> from the board — these are the areas that'll become <u>copper tracks</u>.

6) Finally, any remaining <u>photosensitive layer</u> is <u>stripped off</u>, exposing the bare copper tracks. Now the board is ready for <u>components</u> to be <u>added</u>...

PCBs can be Assembled Manually or Using a Machine

1) <u>Pick and place assembly</u> is the process of <u>placing components</u> onto a <u>PCB</u> in the <u>correct position</u> and <u>orientation</u> — so they're ready to be soldered onto the board.

2) This can be done using a <u>pick and place CAM machine</u> or <u>by hand</u>:

Pick and Place Machines

1) Pick and place machines use <u>suction cups</u> to <u>pick up</u> components and <u>move them</u> to their <u>correct location</u>. The cups <u>rotate</u> so the components are the <u>right way round</u> before being placed onto the board.

2) Machines need to be <u>programmed</u> with the <u>rotation angle</u> and the <u>location</u> of <u>every component</u> on the PCB. This can be a <u>lengthy</u> process.

3) These machines are much <u>quicker</u>, more <u>precise</u> and more <u>consistent</u> than assembling PCBs manually. In <u>mass production</u>, this <u>makes up</u> for the <u>high cost</u> of the <u>machines</u>.

4) Machines are normally only used for <u>surface-mounted</u> components.

Manual Assembly

1) Placing the components by hand can be a <u>fiddly</u> business, especially when components are <u>small</u> — <u>tweezers</u> are often used to <u>increase precision</u> of placement.

2) This is generally used for components with <u>pins</u> that need to be <u>pushed through holes</u>, as machines <u>can't do this</u> very <u>easily</u>.

3) Manual assembly often takes <u>longer</u> than machine assembly (as <u>humans</u> need <u>regular breaks</u>) and is likely to include <u>more mistakes</u> (due to <u>human error</u>).

PCB Production and Surface Treatments

PCB **Lacquering Protects** Components from **Corrosion**

1) A <u>PCB lacquer</u> is a <u>surface treatment</u> for PCBs — it's a <u>thin polymer film</u> that provides a <u>protective barrier</u> against <u>moisture</u>, <u>chemicals</u>, large <u>temperature changes</u> and <u>dust</u>.

2) This is important as <u>electronic components</u> often need to <u>cope</u> with <u>harsh environments</u> — it helps to <u>decrease corrosion</u> (see p.92) and increase the <u>durability</u> of the board.

3) PCB lacquers can be <u>sprayed</u> on, <u>painted</u> on or applied by <u>dipping</u> the whole board into the lacquer.

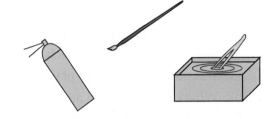

4) Many lacquers are <u>fluorescent</u> — they glow in <u>UV light</u>. This is used in <u>quality control</u> (see p.49), as it makes it <u>easy to see</u> where a coating has been applied and more importantly, if any <u>places</u> have been <u>missed</u>.

Lubricants **Reduce Friction** Between **Moving** Parts

1) <u>Friction</u> is the <u>force</u> that <u>stops</u> two things from <u>sliding past</u> each other. They need to be <u>touching</u> for friction to act. For example, when a <u>bike brakes</u>, the <u>brake pads</u> come into <u>contact</u> with the wheels and the friction between them eventually <u>stops the wheels from turning</u>.

2) Friction can cause problems in a lot of mechanical systems:
   - It can <u>reduce</u> the <u>efficiency</u> of the system — <u>moving</u> parts that are <u>linked</u> or <u>interlock</u> are <u>slowed down</u>, causing <u>kinetic energy</u> (movement) to be <u>partially converted</u> to <u>waste heat energy</u>.

     In an efficient system, only a small amount of the energy that's released goes to waste (e.g. waste heat energy) — the rest is useful.
   - It can lead to <u>interlocking parts</u> being <u>worn down</u>, <u>increasing</u> the <u>chance</u> of them becoming <u>damaged</u> and <u>reducing</u> their <u>lifespan</u>.

3) Mechanical systems are therefore often <u>lubricated</u> to <u>reduce friction</u> — a <u>slippery</u> material called a <u>lubricant</u> is applied <u>between</u> the <u>two surfaces</u> to allow them to slip past each other easily.

4) Importantly though, lubricants <u>don't get in the way</u> of the <u>mechanism</u> — for example, gears will <u>still interlock and work as normal</u>, but just more efficiently and with less wear and tear.

There are <u>loads of types</u> of lubricant for all kinds of <u>applications</u>:
   - <u>Graphite</u> (the material used for pencil leads) is used to lubricate <u>locks</u>.
   - <u>Oil</u> is used on <u>bike chains</u> and in <u>vehicle engines</u>.
   - <u>Silicone</u> or <u>grease</u> is often used to lubricate <u>door hinges</u> and stop them from <u>squeaking</u>.

Lacquering makes PCBs more durable so they last longer...

It's easy to get words like lacquer and lubricant muddled up, especially when you're under pressure in the exam. Remember, a lacquer is a coating that protects PCBs from things like moisture in the air. A lubricant is a material that allows two surfaces to slide past each other easily.

Warm-Up and Worked Exam Questions

Alright, that's everything covered in Section Seven — as always, try the warm-up questions to start and check that the worked exam questions make sense, then delve into the exam questions on the next page.

Warm-Up Questions

1) Steel is often used in car bodies as it is tough and designed to crumple.
   Why are these important properties for a car body to have?
2) What are current and voltage ratings?
3) State the function of the pins and the casing of a Dual in Line package (DIL).
4) Solder is a good electrical conductor. Why is this an important property for solder to have when it is used to attach components to a printed circuit board (PCB)?
5) What is meant by pick and place assembly?
6) a) What type of material are PCB lacquers made from?
   b) What is the function of PCB lacquers?

Worked Exam Questions

1 Oil is an example of a lubricant. It is often used in mechanical systems such as car engines.

a) What is the purpose of a lubricant?

A lubricant allows surfaces that are touching to slip past each other easily.

> You could have also said that a lubricant reduces the friction in a mechanical system.

[1 mark]

b) Give **two** benefits of using a lubricant in a mechanical system.

1. It allows the system to operate with a higher efficiency.

2. It reduces the rate at which interlocking parts are worn down.

> Another valid answer you could have given is that it reduces the chance of moving parts being damaged.

[2 marks]

2 Some of the components in domestic appliances are frequently exposed to water. For example the inside of the dishwasher, shown in **Figure 1**, comes into contact with water during use.

Name **one** suitable material that parts that are exposed to water could be made of. Give a reason for your answer.

Stainless steel would be a suitable material because it is resistant to corrosion.

Figure 1

> Exposure to water causes many materials to corrode, so corrosion-resistance is an important property for materials inside a dishwasher to have.

> Plastic would also have been an acceptable material to give.

[2 marks]

Exam Questions

1   The E12 series are fixed resistors.

a) The resistance of an E12 resistor is stated as **56** ohms **±10%**.
    Give the range its actual resistance lies within.

    ....

    ....

    ....

    ....
                                                                        [3 marks]

b) Coloured bands are used to identify different E12 resistors.
    The numbers these bands represent are shown in **Figure 2**.
    State the colours of the first **three** bands that would be used
    to identify an E12 resistor with a resistance of **68 000** ohms.
    The colours must be stated in the correct order.

    ....

    ....

    ....
                                                            [1 mark]

Colour of band	Number
Black	0
Brown	1
Red	2
Orange	3
Yellow	4
Green	5
Blue	6
Violet	7
Grey	8
White	9

Figure 2

2   Automated soldering methods are used in the mass production of printed circuit boards (PCBs).

Name **one** automated soldering method.

Name: ....

Use sketches and/or notes to give a detailed description of this method.

                                                                        [5 marks]

Revision Questions for Section Seven

Section Seven — short but sweet. Have a go at these questions to test how much you can remember...
- Try these questions and tick off each one when you get it right.
- When you've done all the questions for a topic and are completely happy with it, tick off the topic.

Properties of Components in Systems (p.112-113) ☐

1) a) What is a photosensitive material?
   b) What is used to modify the photosensitive material during PCB production? ☑

2) Explain why using aluminium in motor vehicles can be better than using steel.
   Your answer should include two different properties of aluminium. ☑

3) Suggest one property of rubber that makes it useful in washing machine drive belts. ☑

4) Name the process that makes the surface of aluminium harder. ☑

5) Give one property that you would expect a nickel-chromium alloy to have that makes it
   suitable for use as a heating element in a kettle. Explain your answer. ☑

Standard Components in Systems (p.114-115) ☐

6) What is an integrated circuit (IC)? ☑

7) What is the role of a fixed resistor in an electrical circuit. ☑

8) Work out the resistance and tolerance of a resistor with bands that are coloured
   in the order: orange, white, brown, and silver. ☑

9) What does DIL stand for? ☑

10) Suggest one advantage of a PIC having flash memory. ☑

11) Springs can be sold in varying lengths. Suggest two other features that springs can be sold by. ☑

12) Give two specific characteristics that different types of gears can be sold by. ☑

Cutting, Drilling and Soldering (p.116-117) ☐

13) Name three materials that laser cutters can cut. ☐

14) Suggest a downside of using laser cutters. ☐

15) a) Why are holes made in PCBs?
    b) Name the piece of equipment used for this purpose. ☐

16) When manually soldering, suggest a piece of equipment you would use to melt the solder? ☐

PCB Production and Surface Treatment (p.118-119) ☐

17) a) Describe how a PCB mask is made.
    b) Describe how the mask is used in the photo-etching process of PCBs. ☐

18) Karthik is looking to assemble 10 identical printed circuit boards on a tight budget.
    He's using components that need to be pushed through holes.
    Describe why he should assemble the PCBs by hand rather than using a machine. ☑

19) a) Give two things that a PCB lacquer protects against.
    b) Some lacquers are fluorescent. This allows the lacquer to be viewed under UV light. Describe
        how using fluorescent lacquers is useful in the quality control of PCBs during their manufacture. ☑

20) Give one example of a lubricant. ☐

21) Give two examples of mechanical systems that may need lubrication. ☑

Looking at the Work of Designers

You have to complete a <u>design and make task</u> as part of your <u>non-exam assessment (NEA)</u> (see p.156), but there's also a section of the <u>exam</u> dedicated to <u>designing and making principles</u>. So now seems like a good time to get clued up on all things design — there's lots to consider. First up — looking at the work of <u>existing designers</u>...

Looking at the **Work of Others** Can Give You **Ideas**

1) Being <u>influenced</u> by <u>good designers</u> is a great way to get <u>ideas</u>.
You <u>can't</u> just <u>copy</u> their stuff though — that's <u>illegal</u>.

2) You need to know about the work of <u>at least two</u> of the <u>designers</u> listed below and on the next page.

3) They've all <u>developed styles</u> that have influenced other designers. <u>Investigating</u>, <u>analysing</u> and <u>evaluating</u> the work of some of these designers could influence <u>your designs</u> too.

4) Here's a <u>bit of information</u> on each designer to get you <u>started</u>...

<u>HARRY BECK</u> — <u>(1902-1974)</u> — redesigned the <u>London Underground map</u> in the 1930s. Its <u>simplified layout</u> made it a huge success — maps of many <u>other transport networks</u> now use Beck's style.

<u>DAME VIVIENNE WESTWOOD</u> — <u>(1941-Present)</u> — her <u>iconic clothing</u> became very popular during the <u>punk rock movement</u> in the <u>1970s</u>. She has since become a world famous fashion designer. Her designs often take inspiration from <u>traditional British clothing</u> and <u>historical paintings</u>.

<u>MARCEL BREUER</u> — <u>(1902-1981)</u> — a <u>modernist architect</u> and <u>furniture designer</u>. Some of his best known work includes <u>furniture</u> made from <u>tubular steel</u> and <u>buildings</u> which look like they've been sculpted from <u>concrete</u>.

Modernist design is simple with no ornamentation — it focuses on the function of a product rather than its appearance. Modernist buildings were typically made from concrete, steel and glass.

<u>DAME MARY QUANT</u> — <u>(1934-Present)</u> — a <u>fashion designer</u> who popularised the <u>mini skirt</u>, <u>hot pants</u> and <u>PVC</u> in the <u>sixties</u>. Her clothing often featured <u>white collars</u>, <u>simple shapes</u> and <u>bold colours</u>.

<u>NORMAN FOSTER</u> — <u>(1935-Present)</u> — an <u>architect</u> whose buildings are often constructed from <u>glass</u> and <u>steel</u>. Foster's work includes <u>famous landmarks</u> such as <u>Wembley Stadium</u>, the <u>Millennium Bridge</u> and 'the <u>gherkin</u>' in London.

<u>WILLIAM MORRIS</u> — <u>(1834-1896)</u> — a <u>designer</u> who is best known for his <u>wallpaper</u>, <u>furniture</u> and <u>furnishings</u>. His designs were often based on <u>patterns</u> found in <u>nature</u>.

<u>ALEXANDER MCQUEEN</u> — <u>(1969-2010)</u> — an influential <u>fashion designer</u> known for his <u>theatrical</u>, <u>well tailored</u> clothing and <u>dramatic catwalk presentations</u> displaying his collections.

<u>ALDO ROSSI</u> — <u>(1931-1997)</u> — an <u>architect</u> who also published work on <u>architectural theory</u>. He designed a number of <u>products</u> too, including some for <u>Alessi</u> (see the next page).

<u>PHILIPPE STARCK</u> — <u>(1949-Present)</u> — an <u>architect</u> and <u>product designer</u>. He has designed many products including <u>furniture</u>, <u>kitchenware</u> and <u>vehicles</u>. One of his best known products is his lemon squeezer for <u>Alessi</u>.

Art Nouveau designs are flowing and curvy. They often use floral or insect motifs.

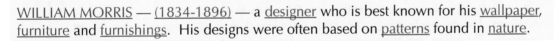

<u>LOUIS COMFORT TIFFANY</u> — <u>(1848-1933)</u> — an <u>Art Nouveau</u> <u>jewellery</u> and <u>glassware designer</u>. He is best known for his work with <u>leaded glass</u>, which he used to produce <u>windows</u> and <u>lampshades</u>.

The Art Deco design style involves bold colours, geometric, zigzag and stepped shapes, bold sweeping curves and sunburst motifs.

<u>RAYMOND TEMPLIER</u> — <u>(1891-1968)</u> — designed <u>Art Deco jewellery</u>. His <u>geometric</u> pieces were different to the traditional styles of the time.

<u>COCO CHANEL</u> — <u>(1883-1971)</u> — a <u>fashion designer</u> known for introducing <u>practical</u>, <u>casual-chic</u> clothing for women, who had traditionally worn corsets and long skirts.

Looking at the Work of Designers

<u>CHARLES RENNIE MACKINTOSH</u> — <u>(1868-1928)</u> — an <u>architect</u> and <u>designer</u>. His work gained most recognition <u>after his death</u>, but his unique style is shown in a number of buildings of his design in <u>Scotland</u>, including the <u>Glasgow School of Art</u>.

<u>GERRIT RIETVELD</u> — <u>(1888-1964)</u> — an <u>architect</u> and <u>designer</u> involved in <u>De Stijl</u> (Dutch modernism). De Stijl designs were basic — they used <u>simple shapes</u>, <u>vertical</u> and <u>horizontal lines</u> and <u>primary colours</u>.

<u>SIR ALEC ISSIGONIS</u> — <u>(1906-1988)</u> — a <u>car designer</u> who designed the <u>Mini</u> in the 1950s and the <u>Morris Minor</u> in the 1940s.

<u>ETTORE SOTTSASS</u> — <u>(1917-2007)</u> — an <u>architect</u> and <u>product designer</u>. He was the founder of the <u>Memphis movement</u>. Memphis designs used many <u>bright</u>, <u>contrasting colours</u> as shown by the Carlton room divider, designed by Sottsass. He also designed <u>Alessi</u> products.

The Memphis movement was at the height of postmodernism. Postmodernist designers were sick of modernist design and thought that style should be the starting point of a product.

You need to investigate the work of <u>at least two</u> of these <u>companies</u> too...

Braun	1)	<u>Braun</u> is an <u>electrical appliance</u> company based in <u>Germany</u>.
	2)	<u>Dieter Rams</u>, who oversaw design at <u>Braun</u> for <u>over 30 years</u>, worked on the basis of '<u>less, but better</u>' — his designs were <u>simple</u>, with <u>no unnecessary details</u>. This made Braun's products <u>easy to use</u> and gave them an <u>industrial</u> look.
Dyson™	1)	Sir James Dyson invented the <u>first bagless vacuum cleaner</u> in the 1980s. Before this, vacuum cleaners <u>lost suction</u> over time and contained <u>bags</u> to collect dust which needed <u>replacing</u>.
	2)	Dyson's design made use of <u>cyclones</u> and a <u>dust bin</u> (a plastic container where dust collects). This gave <u>greater</u>, <u>consistent suction</u> and there was <u>no need</u> to keep buying bags.
	3)	<u>Dyson™</u> have since created a variety of vacuum cleaners designed with a focus on being <u>efficient</u> and <u>innovative</u>. The company also produces a range of other products.
Apple®	1)	<u>Apple®</u> products are instantly recognisable due to their <u>smooth</u>, <u>rounded edges</u> and <u>simple</u> and <u>sleek</u> style.
	2)	Their products are very <u>intuitive</u>, making them <u>user friendly</u>.
Alessi	1)	<u>Alberto Alessi</u> took over the running of his family's kitchenware company in the 1970s. He <u>didn't design</u> any products himself — he employed designers and architects to come up with <u>fun</u> and <u>creative</u> designs which were then manufactured by his company.
	2)	A key part of his idea was to <u>mass-produce</u> products but keep them as <u>stylish</u> and <u>original</u> as possible. The designs were always <u>distinctive</u> and often <u>colourful</u>. A well known example is <u>Philippe Starck's</u> lemon squeezer.
Gap	1)	<u>Gap</u> is an <u>American clothing company</u> which sells <u>casual</u>, <u>everyday</u> wear (e.g. jeans, t-shirts, etc.).
	2)	The company started out in 1969 as <u>single shop</u>, initially just selling <u>jeans</u> alongside <u>records</u> (to attract <u>young people</u> into the shop). Its <u>popularity</u> led to the company quickly increasing its <u>number of shops</u> and <u>range of clothing</u>. The company now has thousands of stores <u>worldwide</u>.
ZARA	1)	<u>Zara</u> is a <u>Spanish clothing company</u> which sells <u>up-to-date</u>, <u>fashionable</u> clothes at <u>affordable</u> prices.
	2)	The company <u>constantly</u> responds to <u>evolving fashion trends</u> and <u>customer feedback</u>. This means their clothing ranges <u>change frequently</u>, keeping their <u>customers coming back</u> to see what's new.
Primark®	1)	An <u>Irish clothing company</u> which opened its first store in <u>Dublin</u> (under the name of <u>Penneys</u>) in 1969. <u>Primark®</u> now has stores <u>across the UK</u>, in <u>Europe</u> and the <u>USA</u>.
	2)	Primark® aims to provide <u>current trends</u> in fashion at a <u>low cost</u> to its customers.
Under Armour®	1)	<u>Under Armour®</u> is an <u>American sportswear company</u>. It was founded by <u>Kevin Plank</u>, an <u>American footballer</u>. He created a shirt that <u>wicked sweat away</u> from the <u>skin</u> after becoming fed up with his sportswear <u>absorbing</u> large amounts of <u>sweat</u> during exercise.
	2)	His shirts were <u>very popular</u> with other football players, leading to the company expanding. <u>Under Armour®</u> now sell a <u>wide range</u> of <u>sportswear</u> and <u>fitness devices</u>.

You can get inspiration for your own designs from the work of others...

You don't need to memorise every designer and company — but you should know about two of each.

Understanding User Needs

Different people have different needs (they also have different values — see p.11).
These needs must be taken into account when designing a product for a particular group of people...

Ergonomics Means Making the Product Fit the User

Designing a product so that it's comfortable and easy to use is known as ergonomics (or human factors).

1) Products need to be designed so that their size and proportions fit
the needs of the user. For example, the buttons on a calculator
need to be big enough for the user to press them individually.
Textile products need to vary in size (e.g. clothes) or be adjustable
(e.g. rucksacks) to fit the user.

2) Ergonomic design also helps to ensure that using the product won't
cause health problems. For example, an ergonomically designed chair
should prevent you suffering backache from using the chair regularly.

3) Ergonomics needs to take the target market into account, e.g. a chair
for a five-year-old needs to be a different size from a chair for a
fifteen-year-old (obviously).

4) Designers use body measurement data (also called anthropometric data
— see below) to make sure the product is the right size and shape.

Many chair designs achieve this by allowing you to have your feet on the floor with your knees at a right angle and your back supported.

Anthropometric Data are Measurements of Humans

1) Measurements of human body parts are called anthropometric data. This includes things like:

- Upper arm length
- Head circumference
- Height (stature)
- Chest depth
- Shoulder height
- Hand width
- Knee height
- Little toe length
- etc...

2) These measurements are collected from a wide range of people with different body sizes.

3) To make products of the right size, designers need to know the
likely body measurements of the intended users...

1) First, they work out what measurements they need. For example, if their product is a novelty
mask it doesn't matter how long the users' legs are — they only need to know about their heads.

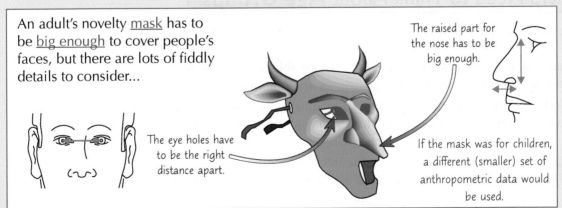

An adult's novelty mask has to be big enough to cover people's faces, but there are lots of fiddly details to consider...

The eye holes have to be the right distance apart.

The raised part for the nose has to be big enough.

If the mask was for children, a different (smaller) set of anthropometric data would be used.

2) Next, they find out what these measurements are for the typical user of the product — these
measurements will have been collected by sampling lots of people from the target group.

3) Then, they design the product to fit a range of users within the target group (e.g. adults for
the novelty mask). There's more on how this is done on the next page...

Understanding User Needs

Products are Often **Designed** to **Fit 90%** of the **Target Market**

1) Manufacturers try to design products so that they'll be suitable for <u>most people</u> in the <u>target market</u> — they often aim to make the product fit <u>okay</u> for <u>90%</u> of the target users.

2) For example, a manufacturer designing an office <u>desk</u> might look at data for the <u>heights</u> of a sample of adults aged between 20 and 65:

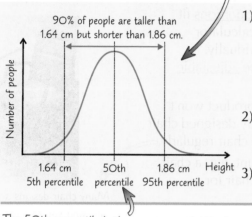

90% of people are taller than 1.64 cm but shorter than 1.86 cm.

Number of people

1.64 cm — 5th percentile
50th percentile
1.86 cm — 95th percentile
Height

1) Manufacturers often try to make 'one size fits all' products, but actually, it's 'one size fits 90%' — they use the <u>5th and 95th percentiles</u> as cut-off points:

> <u>5th percentile</u>: this means that <u>5%</u> of people are <u>shorter</u> than this value.
> <u>95th percentile</u>: this means that <u>95%</u> of people are <u>shorter</u> than this value.

2) Here, this means the <u>bottom 5%</u> (very short people) and the <u>top 5%</u> (very tall people) <u>aren't catered for</u>. A one-size product wouldn't suit either extreme.

3) These people would probably have to use a specialist supplier or have products <u>custom-made</u> for them.

The 50th percentile is the average — 50% of people are shorter and 50% of people are taller.

Products Need to be **Accessible** for **Disabled Users**

Lots of products are specifically designed to help people with <u>disabilities</u>.

1) Some packaging (e.g. for medication) has <u>Braille</u> labelling to give blind people information.

2) Control buttons can be made <u>brightly coloured</u> and <u>extra large</u>, so they're easy to find and press. For example, telephones, TV remotes and calculators can be made with big buttons.

3) Products such as smoke alarms can be designed with <u>visible</u> signals as well as audible ones so that deaf people can be alerted to fires.

4) Designers also have to think about <u>wheelchair users</u>. For example, trains and buses need to be designed to have wheelchair access.

Designers Need to Think About **Age Groups**

People in different <u>age groups</u> have different <u>physical limitations</u>. For example:

1) Small children and elderly people may not be able to manipulate <u>small parts</u> and may have difficulty undoing <u>fastenings</u> and opening <u>packaging</u>.

2) Elderly and infirm people might have difficulty <u>holding</u> and <u>using</u> products. Designers can think about putting large, easy-to-grip handles on, say, cutlery.

<u>Older people</u> may also have different <u>requirements</u> from <u>textile products</u>, e.g. they may prefer shoes with <u>lower heels</u> and <u>wider fittings</u> for comfort.

Anthropometric just means measurements of humans...

Remember, designers need to think carefully about who a product is aimed at and make sure it's suitable for them. For example, you wouldn't design a child's cup that had sharp edges or was easily breakable.

Design Briefs and Specifications

The process of designing and making something is called the <u>design process</u> (gosh).

Designing Starts with the **Design Brief**

So, someone gets an idea for a <u>new product</u>.
They decide to <u>employ a designer</u> to work on the idea.

1) The person who hires the designer is called the <u>client</u>.

2) The <u>client</u> gives the designer a <u>design brief</u>...

3) The design brief is a <u>starting point</u> for the development of the product. It should include:

> - what <u>kind</u> of product is needed or wanted (and <u>why</u>)
> - how the product will be <u>used</u>
> - <u>who</u> the product is <u>for</u> (the <u>target market</u>)

4) The designer needs to pick out all the <u>important features</u> of the design brief.

5) One way of doing this is by drawing a <u>spider diagram</u>. It's a quick way to help analyse the problem and to help the designer decide what they need to <u>research</u>.

DESIGN BRIEF FOR
BACK SCRUBBER/SOAP HOLDER

No currently commercially available back scrubber has an in-built capacity for soap storage. We want you to design a product to meet this need for those people with grubby backs and modest soap storage requirements (1 or 2 bars).

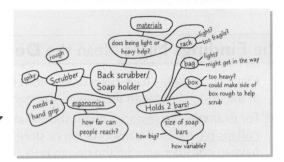

A design doesn't have to be based on the needs or wants of a <u>client</u> — a product idea can be based on <u>your</u> wants, needs and interests too.

Research Helps the Designer Get **Ideas**

The point of doing research is to:

1) check that people will actually <u>want the product</u>.

2) find out what people in the target market <u>like or dislike</u> about similar <u>existing products</u>.

3) find out what <u>materials</u>, standard <u>components</u> and <u>techniques</u> would be suitable for the product, and how they will affect the manufacturing and selling <u>costs</u>.

There are <u>different types</u> of research:

Market Research

<u>Questionnaires</u> or <u>interviews</u> can be used to find out people's likes/dislikes and so on. This helps the designer understand what the <u>target group</u> wants from a product.

There's loads more on carrying out research on pages 129-132.

Product Analysis

Designers can <u>examine</u> a current product or <u>disassemble it</u> (take it <u>apart</u>) to:

- figure out how it's <u>made</u> (e.g. materials, components and processes) and how it <u>works</u>.
- check for <u>good</u> and <u>bad</u> features, including how ergonomic the design is (see page 131).
- <u>collect data</u> on factors that could affect the product's <u>performance</u>, e.g. its <u>size</u>, <u>weight</u> etc. Several products can then be <u>compared</u> at once using <u>tables</u> or by plotting the data on <u>graphs</u>.
- test how it <u>feels</u>, <u>looks</u>, <u>smells</u> or <u>tastes</u> (sensory analysis).

Then they should have a good idea of what <u>materials</u> and <u>techniques</u> they need to research for the product.

Product analysis can identify the pros and cons of existing products...

Designers don't just go straight from the design brief to sketching ideas. They first need to check out other similar products already on sale and get some information about the target market for the new product.

Design Briefs and Specifications

There's loads more on how to actually <u>carry out research</u> on the next few pages.
But before we get onto that, let's take a look at what designers do with their research <u>findings</u>...

Research is Used to Draw Conclusions

Once they've done some <u>market research</u> and <u>product analysis</u>, designers should have loads of information. They use the information to help with their design. They should:

1) <u>Summarise</u> what they've found out — pick out the <u>most important</u> and useful <u>findings</u>. E.g. shower gel is just as popular as soap. This might involve <u>interpreting data</u> and <u>graphs</u>.
2) <u>Explain</u> what <u>impact</u> each finding will have on their design. E.g. the design should be able to hold bottles of shower gel as well as bars of soap.

The Findings Might Mean the Design Brief is Reconsidered

1) Once <u>conclusions</u> have been drawn from the research, the designer might want to <u>look again</u> at the <u>design brief</u> with the client — remember, the design brief is just the <u>starting point</u>.
2) For example, the <u>research</u> might show that most people want the soap holder/back scrubber to be <u>collapsible</u> for easy transportation or store a <u>minimum of 8 bars of soap</u>. As a result, the designer may have to consider <u>altering</u> the <u>design brief</u>, with any <u>decisions</u> made being <u>discussed</u> with the <u>client</u>.

The Design Specification is a List of Conditions to Meet

1) Once the <u>design brief</u> is agreed, a <u>design specification</u> can be made.
2) The design specification gives certain <u>conditions</u> that the product must meet. These conditions (often called <u>design criteria</u>) should take account of the <u>research findings</u>.
3) It's best to write a specification as <u>bullet points</u>. Each point should be <u>explained</u> and show how the <u>research</u> helped the designer to decide on that point. The specification should cover the following things:

- Size — how big it is
- Aesthetics — how it looks
- Consumer — who will buy it
- Function — what it will do
- Quality — e.g. the required finish
- Cost — the price range
- Materials — what it is made of
- Safety — how to make sure it's safe
- Environment — the impact on the world
- Sustainability — its future impact

Tally chart showing the minimum length of back scrubber that people needed to scrub the middle of their back.

Length (mm)	No. of people
150	I
200	IIII
250	IIII III
300	IIII IIII
350	IIII I
400	II
450	

See page 132 for more on why environmental issues need to be considered in your design specification.

Example:

SIZE
- It should weigh 300 g or less, as most people were comfortable with this.
- The minimum length will be 400 mm, as that was long enough for everyone to reach the middle of their back.

AESTHETICS
- It should be multicoloured because most people said they would like the product to be very colourful.
- The handle should be easy to grip but not feel rough, because some people were concerned that a rough surface might hurt their hands after prolonged use.

REVISION TIP

Research can cause the design brief to change...

It's important to know the difference between design briefs and design specifications. A **b**rief is written at the **b**eginning of the development of a product. A specific**a**tion is made **a**fter research has been carried out and the brief agreed — it contains all the points a product should meet.

Market Research

As we said on the previous two pages, market research is important for designing a product that people will want to buy. Now it's time to go into more detail about how exactly market research is carried out...

Identify your Target Group

1) Even the very best products won't be everyone's cup of tea — some people will like them, some won't.

2) You might be given a target market in the design brief, or have a specific target group you want to design for. If not, you need to work out which people are most likely to buy your product — this should be your target group. Ask members of your target group what they want the product to be like.

3) You can group people by things like age, gender, job, hobbies, lifestyle, amount of money they earn, or anything else — it'll probably be a combination of a few of these things.

> For example, if you're trying to sell wide-legged, elasticated-waist, linen trousers, you may decide to target them at middle-aged women who are buying their summer clothes.

> But if you've designed some fetching stripy knee-length socks, you'd be better off targeting them at teenage girls in winter.

Think Carefully About What You Want to Find Out

Once you're clear on exactly who your target group are, you need to decide what to ask them.
You could find out:

1) Some information about the person answering your questions. This could help you make sure they're within your target group, or give you extra information.
   * Are they male or female? (probably best to judge for yourself rather than asking...)
   * What age bracket are they in? (11-15, 16-20, 21-25 etc.)
   * What job or hobbies do they have?

2) Do they already buy the kind of product you're thinking of developing?

3) Do they like a particular style or colour?

4) How much would they be prepared to pay for this kind of product? This could affect your budget — the lower the selling price, the lower the manufacturing costs will need to be to make a profit.

5) Where would they expect to buy it from? Again, this could affect costs. E.g. if the majority of your target group buy cheap items of furniture from IKEA®, you should also design a low-cost item.

6) Is there something they would like from the product that existing products don't have?

> At a later stage in the design process, when you've generated ideas, you may decide to ask the target group more questions.
>
> E.g. will they buy your version of the product? Explain the advantage of your product over existing ones — would that be enough to tempt them to buy your version?

Questionnaires are Forms for People to Fill in

Now, how to phrase those all-important questions. There are two basic types of question:

1) Closed Questions — these have a limited number of possible answers, e.g. 'do you ever use a bag?' can only be answered 'YES' or 'NO'. Analysing is easy, and you can show clear results at the end. Closed questions include multiple-choice questions — these give a choice of answers. Sometimes the person answering can pick more than one. For example:

Q4.	Which of these types of bag do you own?	
Satchel ☐	Clutch ☑	Rucksack ☑

Market Research

2) <u>Open Questions</u> — these have <u>no set answer</u>, e.g. <u>what's your favourite type of bag, and why?</u>
They give people a chance to provide details and opinions. This type of questioning takes more time
and it's harder to draw conclusions from the results. But you could gain valuable information.

Interviews are Face-to-Face Conversations

1) For interviews, you can <u>start off</u> by asking the same sort of questions as in questionnaires
— but then take the chance to ask <u>follow-up</u> questions, based on the answers you get.

2) Get your interviewees to give you <u>extra information</u> to explain their answer — this might help you get
more <u>ideas</u> for your product. E.g. if their favourite type of wood is oak, ask them <u>why</u> they like it.

3) Interviews can give you more <u>detailed</u> information than questionnaires — you can have short
<u>conversations</u> with people you're aiming to sell to. Just make sure you <u>stick to the point</u>.

4) A problem with interviews is that it's often more <u>difficult</u> to <u>analyse</u> the results than with questionnaires
— it's harder to <u>compare opinions</u> when the topics covered vary slightly from person to person.

5) It can be useful to <u>record</u> the interview while it is taking place — audio recordings mean that the
interview can be <u>analysed at a later date</u>, and it means that the person asking questions doesn't need
to <u>stop the flow of conversation</u> to take notes. If a visual recording is also taken, then the person's
<u>body language</u> can also be <u>analysed</u> to help interpret their answers.

Focus Groups Provide a Deeper Level of Feedback

1) Holding a <u>focus group</u> is a bit like carrying out a <u>group interview</u> — the group is asked for their <u>opinions</u>
on either the proposed product <u>in general</u>, or one <u>particular aspect</u> of it such as the <u>packaging</u>.
Like interviews, focus groups are a great way of getting more in-depth <u>feedback</u> from your <u>target group</u>.

2) The group should be made up of people who <u>don't already know each other</u>.
This is important, as people are often more likely to <u>say how they feel</u> with <u>strangers</u>,
whereas people they know may <u>influence</u> their <u>feedback</u>.

3) The aim of the focus group is to <u>generate a guided discussion</u> — where members
can <u>freely</u> give their opinions to questions that have been <u>prepared in advance</u>.

4) Each group should have a fairly <u>small number</u> of <u>people</u> (e.g. 6-10) and be asked a set of <u>open-ended</u>
<u>questions</u>. It's also a good idea to hold <u>more than one</u> focus group so that <u>results can be compared</u>.

ICT can be used to Research, Analyse and Present Information

1) You could write a <u>questionnaire</u> using <u>word processing</u> software — so it'll be <u>neat</u> and easy to read.
This could be posted on a <u>website</u> that your <u>target group</u> is likely to use, so they could fill it in <u>online</u>.

2) You could use the <u>internet</u> to
look up <u>sales figures</u> and collect
them in a <u>spreadsheet</u>.

3) You could use <u>spreadsheets</u> to
<u>organise and sort data</u>
(e.g. your questionnaire results).
Spreadsheets also allow you to
<u>analyse</u> your data and <u>present</u>
it using <u>charts</u> and <u>graphs</u>.

EXAMPLE:

**Jordan is conducting market research for her new shoes, and
has received 35 responses to her questionnaire. 20% of the 35
people that responded said their favourite shoe colour was red.
How many people said their favourite shoe colour was red?**

1% of people = 35 ÷ 100 = 0.35

so 20% of people = 20 × 0.35 = <u>7 people</u>

In a spreadsheet, you
can use a formula to
do this calculation.

Computers make analysing market research data much easier...

Product Analysis

Next up, it's <u>product analysis</u> — this is where you <u>research</u> and <u>analyse</u> some <u>existing products</u> to give you a few <u>ideas</u> on what to use in your own designs. It will help you to write the <u>design specification</u> too...

There's Lots to Consider When **Analysing a Product**

1) As mentioned on p.127, <u>product analysis</u> can be done by looking at the <u>outside</u> of a product or <u>taking it apart</u>. There are <u>several</u> things to consider:

Function

1) Function is what the product is <u>intended</u> to do — its job.
2) <u>Disassembling</u> (taking apart) a product can help you find out <u>how it works</u>.
3) Make careful notes as you disassemble something, and record what <u>component parts</u> have been used and how it's <u>structured</u>, using <u>sketches</u> or <u>photos</u>.

Form

1) This is the <u>shape</u> and <u>look</u> of the thing, e.g. colour, texture and decoration.
2) A product could be <u>old-fashioned</u> or <u>modern-looking</u>. It could have <u>flowing curves</u> or it might be very <u>angular</u> with lots of corners. The general look of a product is also known as its <u>aesthetics</u>.

Ergonomics

1) Ergonomics is about <u>designing</u> products so that their <u>size</u> and <u>proportions</u> fit the users' needs.
2) For example, a <u>hand-held</u> product such as a touchscreen phone needs to <u>fit well</u> in one hand, and many are designed so that <u>all parts</u> of the <u>screen</u> can be <u>reached</u> with your thumb.
3) Designers use <u>body measurement data</u> (<u>anthropometrics</u>) to make sure the product is the <u>right size</u> and <u>shape</u>.

Ergonomics is also covered on page 125.

Competition and cost

1) You need to consider <u>value for money</u>. For example, if you're looking at a <u>hairdryer</u>, find out whether it's cheaper or more expensive than <u>similar</u> hairdryers.
2) You'll also need to look at <u>how it performs</u> compared to these other hairdryers.

Sustainability

<u>How much</u> does <u>making</u> or <u>using</u> the product harm the <u>environment</u>? E.g. most cars emit carbon dioxide (which causes global warming) and <u>various other pollutants</u>.

There's more about sustainability on the next page.

Materials

Product analysis should include looking at <u>what materials</u> have been used, and <u>why those materials</u> were chosen.

Manufacture

1) Consider all the <u>processes</u> that have been used to make the product.
2) This includes things like which <u>techniques</u> were used to <u>shape parts</u> of a product, e.g. the pieces of a garment.
3) Don't forget to check if any parts have been assembled separately and plonked into the product later — the term for that is <u>sub-assembly</u>.

2) It's important that you <u>record</u> the findings of your analysis — <u>make notes</u> so you can remember what you thought, and <u>compare</u> it to other products you look at. You can use a spreadsheet to record data.

3) Once you've considered all these aspects of the product, you should make an <u>evaluation</u> — a <u>judgement</u> about what does or doesn't work about the product, and how <u>effective</u> the different features are.

Form, ergonomics, sustainability — all things to consider...

You probably think this seems a lot to think about before starting on the design of a product. But it's all worth doing. There's no point coming up with a new product if it's no better than an existing one...

Product Analysis

Which **Materials** are Used **Matters**

You need to consider the environmental impacts of materials when doing product analysis:

1) Some materials are toxic, e.g. some paints, varnishes and dyes.

2) Many materials are made from finite resources. For example, most plastics and synthetic fibres are made from fossil fuels (see p.12), which will eventually run out.

3) Products that use recycled materials, e.g. recycled metals and plastics, are more environmentally friendly.

4) It's also better to use sustainable materials, e.g. wood, paper and cotton. Softwoods (which can be regrown in a person's lifetime) are a better choice than hardwoods (which take ages to grow). Wood and paper from ethical sources (e.g. sustainably managed forests) won't contribute to deforestation (see p.58).

5) Many products are thrown away — it's good if these products are made from biodegradable materials (materials that will rot away naturally) or recyclable materials. For example, waste wood is biodegradable and can also be recycled into manufactured boards.

Processes Have **Environmental** and **Social** Impacts Too

You should also think about the processes used to make products:

1) Does the manufacturing process cause pollution? For example, how much waste material will be produced and how will it be disposed of?

2) Does the process use a lot of energy? If manufacturers wanted to be extremely responsible, they'd try to use renewable energy sources like wind power or hydroelectricity.

3) Was the product made under good working conditions? For example, does the manufacturer pay workers fairly, give them protective clothing, etc.

Fairtrade Makes Sure **Producers** get a **Fair Price**

1) A Fairtrade product, or product containing Fairtrade materials (e.g. Fairtrade gold or Fairtrade cotton), can have a smaller social and environmental impact.

2) Fairtrade producers must meet particular standards. This gives benefits to workers, the environment and the producers themselves:

Social standards provide:
* Fair terms of trade (higher prices) for the produce of small-scale farmers and workers, allowing them to earn a better income so they can afford to feed their families.
* Safer working conditions for farmers and workers.
* Fairtrade premiums — money for farmers and workers to help their community in an area they wish to improve, e.g. schools, roads, healthcare etc.

Environmental standards aim to reduce environmental impact in ways such as:
* reducing the use of pesticides
* reducing greenhouse gas emissions
* protecting biodiversity.

3) For these reasons, designers may deliberately choose to use Fairtrade materials in their product over those with fewer social and environmental benefits.

It's important to consider the environmental impacts of a product...

Make sure you learn the social and environmental impacts a product can have. It'll be worth it — honest.

Warm-Up and Worked Exam Questions

Designers need to take a lot of information on board before they even begin to design the product itself. It's a bit like revising for an exam. Check you've taken everything in by having a go at these questions.

Warm-Up Questions

1) Suggest three types of anthropometric data that might be needed when designing a football shirt.
2) What is the name given to a person who hires a designer?
3) What is a design specification?

Worked Exam Questions

1 Designs are often aimed at specific target markets.

a) Target markets can be defined by age or lifestyle.
Give **two** other ways in which a target market can be defined.

1. Gender ....
   There are many different answers you could have given here — flick back to p.129 for other ways
2. Job ....
   in which a target market can be defined.

[2 marks]

b) Identify a possible target market for the pin-striped suit shown in **Figure 1**.
Explain your answer.

Male professionals, because it's a man's suit and is most ....

likely to appeal to men who wear suits every day to work. ....

[2 marks]

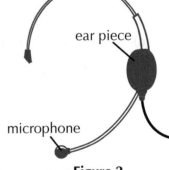

Figure 1

2 **Figure 2** shows a design for an adjustable headset.

a) Suggest **two** pieces of anthropometric data that should be considered when designing the headset.

1. The distance between the ear and mouth. ....

   ....

2. The vertical distance between the ear and the top of ....

   the head. *The width of the head would also be a valid answer here.* ....

[2 marks]

b) The headset is only available in one size. Although it is adjustable, it will not fit a user that falls into the 97th percentile for head size. Explain why.

Products that are only available in one size are often designed to fit 90% of the ....

target users, using the 5th and 95th percentiles as cut-offs. A person in the 97th ....

percentile for head size will have head dimensions larger than the range catered for ....

by the headset, so it won't fit their head. ....

[2 marks]

Exam Questions

1   A company are designing a low cost, winter coat for women, made from environmentally-friendly materials. **Figure 3** shows an example of a women's winter coat already on the market.

Analyse the product shown in **Figure 3** in terms of:

You may need to use your knowledge on materials to help you answer these questions.

a)   its functionality

....

....

....

....

....

[2 marks]

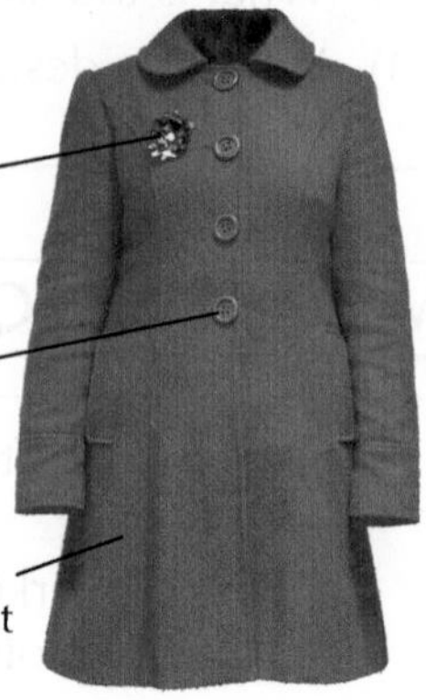

metal brooch

recycled plastic buttons

nylon with a water-resistant finish

Figure 3

b)   the cost of its materials

....

....

....

[2 marks]

c)   the sustainability of its materials

....

....

....

[2 marks]

d)   The company are also trying to minimise the negative social impact of their garments.

State **one** negative social impact that they could try and avoid during the coat's manufacture.

....

[1 mark]

2   Compare the work of **two** of the designers listed below.

In your answer you should try to include what each designer is well-known for, the style(s) they work with, and examples of their work.

- Harry Beck
- Vivienne Westwood
- Marcel Breuer
- Mary Quant
- Norman Foster
- William Morris
- Alexander McQueen
- Aldo Rossi
- Philippe Starck
- Louis Comfort Tiffany
- Raymond Templier
- Coco Chanel
- Charles Rennie Mackintosh
- Gerrit Rietveld
- Sir Alec Issigonis
- Ettore Sottsass

Remember, this question is all about comparing two designers, so you'll need to focus on their similarities and differences.

[8 marks]

Design Strategies

You can use <u>design strategies</u> to come up with <u>initial design ideas</u> without you <u>getting stuck</u> on a <u>bad one</u>...

There are **Several** Different **Design Strategies**

Designing is a really <u>complex</u> process and there are several different ways of doing it:

Systems Approach

This means <u>breaking down</u> the <u>design process</u> into a number of <u>different stages</u> and doing each in turn — e.g. writing the <u>design specification</u>, coming up with <u>ideas</u>, <u>developing</u> your ideas, <u>research analysis</u>, etc. This is a very <u>orderly</u> and <u>reliable</u> method of designing.

User-Centred Design

In this strategy, the <u>wants</u> and <u>needs</u> of the <u>user</u> are <u>prioritised</u> — their <u>thoughts</u> are given lots of <u>attention</u> at <u>each and every stage</u> of the design process.

Iterative Design

This is the design strategy that you'll use to <u>make</u> your <u>prototype</u> during your <u>NEA</u> (non-exam assessment). It is centred around a <u>constant process</u> of <u>evaluation</u> and <u>improvement</u> — see below.

A prototype is a test version of your design — see p.147 for more about prototypes.

Iterative Design is a **Circular Process**

1) The iterative design process <u>doesn't stop</u> once you've come up with your <u>first good design</u> and made a <u>prototype</u>. It still needs to be <u>tested</u>, <u>evaluated</u> and <u>improved</u> based on the <u>results of testing</u>.

2) The whole process is then <u>repeated</u> for the <u>new</u> and <u>improved prototype</u>, so it can get pretty <u>circular</u>...

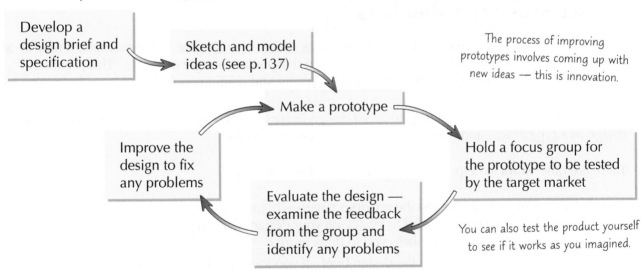

Develop a design brief and specification → Sketch and model ideas (see p.137) → Make a prototype → Hold a focus group for the prototype to be tested by the target market → Evaluate the design — examine the feedback from the group and identify any problems → Improve the design to fix any problems → (back to Make a prototype)

The process of improving prototypes involves coming up with new ideas — this is innovation.

You can also test the product yourself to see if it works as you imagined.

3) This process is <u>repeated</u> until all the <u>problems</u> have been <u>identified</u> and then <u>fixed</u>. Then the product is ready for production.

You can Consult **Other People** During the Design Process

1) <u>Collaboration</u> is all about working with other people. It allows you to gain <u>specialist knowledge</u> from <u>across</u> a variety of <u>subject areas</u>. It's often a <u>key</u> part of the <u>design process</u>, and is involved in <u>all</u> of the <u>strategies</u> mentioned above.

Iterative design works to constantly refine a prototype...

You need to know about the different approaches to design. You'll use the iterative one during your NEA.

Design Strategies

2) You'll need to <u>work with different groups</u> of people to make sure your design is the best it can be.

Client Involvement

1) If you're working with a client, you may find it helpful to show them your <u>designs</u>, <u>models</u> and <u>prototypes</u> early on in the design process, to make sure it fits the <u>image</u> they had in mind.

2) If it's <u>not</u> what the <u>client</u> imagined, it offers them the opportunity to give <u>feedback</u> to <u>improve future iterations</u> of the design.

3) Each client will be <u>different</u>. Some may want to take a <u>hands-on approach</u> and be heavily involved in the design process, whereas others may <u>collaborate less</u> and leave it more to the designers.

User Involvement

1) This involves asking a <u>sample</u> of the <u>target market</u> for input in the design process.

2) The aim is to get feedback from <u>potential users</u> on your <u>design ideas</u>, <u>models</u> and <u>prototypes</u>, and make improvements so that the end product is more <u>appealing</u> to the <u>target market</u>.

Expert Opinions

1) Experts are other <u>professionals in the industry</u>, e.g. seamstresses, electronics experts etc.

2) You can <u>benefit</u> from their <u>experience</u> by asking them to look at your design and <u>suggest improvements</u>. For example, if you're making an <u>evening dress</u>, a seamstress may be able to suggest ways of altering a pattern so it's <u>cheaper</u> and <u>easier to make</u>, but still <u>looks good</u>.

Don't Get **Stuck** on a **Bad Idea...**

1) When you're designing a product it's easy to get <u>stuck on a particular idea</u> — for example, it may be an idea that is <u>similar</u> to an <u>existing product</u>, or a design that you've thought of <u>before</u>. This is called <u>design fixation</u> — it can stop you from thinking <u>creatively</u> and coming up with an <u>innovative design</u> idea.

2) <u>Following</u> a <u>design strategy</u> can help to <u>avoid design fixation</u> — they encourage you to look at your design in a <u>critical way</u> and make <u>improvements</u> where necessary.

3) There are other ways to help <u>avoid</u> design fixation. These include:

- <u>Collaboration</u> (see above and the previous page) — <u>brainstorming ideas</u> with <u>other people</u> and getting <u>honest feedback</u> on your design ideas might allow the <u>fixation</u> to be <u>broken</u>. It may help you to think more <u>creatively</u> too.

- <u>Focusing on new solutions</u> to the design brief. If you focus too much on existing products, it can be easy to attempt to <u>copy</u> them. It's great to use them as <u>inspiration</u>, but your aim should be to create a <u>new</u> and <u>unique</u> product, so make sure your ideas take a <u>fresh approach</u> to your product's <u>specific requirements</u>.

Design strategies usually involve collaboration with others...

It's really important you get your head around the different design strategies that can be used. To make sure it's clear in your mind, write a brief paragraph describing each of the three main design strategies — systems approach, user-centred design and iterative design.

Exploring and Developing a Design Idea

The first part of the <u>iterative design process</u> (see p.135) is <u>developing</u> a <u>design</u>. This is all about going from some <u>roughly sketched ideas</u> to your <u>first design</u> that you want to be made into a <u>prototype</u> (see p.147).

Detailed Sketches Help You Work on Finer Points of the Design

1) Your <u>initial sketches</u> will probably have been <u>rough</u>, <u>freehand</u> pencil drawings.

2) Trying out some <u>more detailed sketches</u> is the next stage.

3) It helps you to see what will actually <u>work</u> in practice and it might help you decide on <u>details</u> you hadn't thought about before, e.g. the sizes or positions of components, or how parts should be <u>constructed</u> and <u>fitted together</u>.

See pages 139-142 for lots of different drawing techniques.

Use Modelling to Improve Your Design

Models are often scaled-down (less than full-size) versions of a design.

1) Modelling is just making <u>practice versions</u> of your design, or parts of your design. It's a good way to <u>visualise</u> your design in <u>3D</u>, and to spot (and solve) any <u>problems</u>.

2) The modelling stage should also be used to try out <u>different materials</u> and <u>joining techniques</u>, and think about whether you could <u>reduce</u> the number of <u>parts</u> to make construction easier.

3) You can make models using materials that are <u>easy</u> and <u>quick</u> to <u>work with</u>, e.g. <u>cardboard</u>, <u>balsa</u>, <u>jelutong</u> (an easily workable wood), <u>newspaper</u> (used in textiles) or <u>foam core board</u>.

4) You can also use <u>construction kits</u> — these have different sized and shaped parts for you to build with.

5) Some types of modelling are only used for a <u>specific type</u> of <u>product</u>:

Toile

1) <u>Toiles</u> are <u>early versions</u> of a <u>clothing</u> design. They can be <u>worn</u> by a <u>model</u> or put on a <u>mannequin</u>, and they're used to work out the <u>proportions</u> and <u>fit</u> of the garment. Toiles for <u>one-off garments</u> are often fitted to the <u>intended wearer</u>.

2) The <u>fabric</u> used to make the toile should:
- Be <u>cheap</u> — so the design can be <u>experimented</u> with and <u>improved</u>, without costing too much and <u>wasting</u> the <u>real fabric</u> you're planning to use.
- Be <u>lightly coloured</u> — so you can <u>mark up</u> any <u>changes</u> needed onto it.
- Have <u>similar properties</u> (e.g. weight, stretchiness, etc.) to the <u>real fabric</u> you're planning to use — this makes sure the <u>drape</u> of the garment can be <u>perfected</u> too, as it will be <u>similar</u> to when the real fabric is used.

3) <u>Calico</u> (an <u>unbleached cotton</u>) is a commonly used toile fabric.

Breadboards

1) Breadboards are used to <u>test</u> whether a <u>circuit design works</u>.

2) They are <u>boards</u> with <u>rows of holes</u> that <u>electronic components</u> can be <u>pushed through</u>. <u>Wires</u> are plugged into other holes to <u>complete the circuit</u> for testing.

3) Breadboards are useful as they <u>don't require soldering</u> (see p.117) — this is important when <u>developing</u> a circuit, as components can <u>easily be removed</u> and the <u>circuit easily changed</u>.

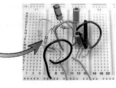

6) <u>CAD/CAM</u> (see p.4-5) can be used for <u>modelling</u> too. CAD can be used to draw a <u>detailed 3D design</u> and then <u>CAM machines</u> such as <u>3D printers</u> can use this design to produce a <u>model</u>.

7) <u>Mathematical modelling</u> is another way to model. It can show how an <u>object</u> or <u>system</u> will behave in <u>reality</u>. These models are made using <u>data</u> and information about <u>known relationships</u> between variables (factors that change).

Mathematical models can display information in a numerical or graphical form.

Modelling can show you if certain design features are feasible...

There are loads of different ways to make models of your design for testing and evaluating, but they all have a few things in common. The materials must be easy and quick to work with, and the model must be functionally and/or aesthetically similar to the final product so the results from testing are realistic and useful.

Exploring and Developing a Design Idea

Test and Evaluate Each Model

1) After you've made the first model, do some <u>tests</u> to check that it's how it should be.

2) You'll probably find there are some things that <u>don't work out</u> quite how you'd hoped:

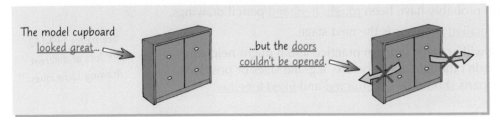

The model cupboard <u>looked great</u>.... ...but the <u>doors</u> <u>couldn't be opened</u>.

3) <u>Write down</u> what the problem is, and suggest how to <u>fix it</u>, e.g. use a different type of hinge. It might be a good idea to <u>mark up</u> the <u>changes</u> required <u>on the model itself</u> (e.g. as you would with a toile).

4) Record how the design develops — <u>take photos</u> of your models.

5) You should also <u>evaluate</u> each model against the <u>design specification</u> (see p.128). Take each point on the specification and see if your model is up to scratch.

6) You might end up having to <u>modify</u> your <u>specification</u> after you've evaluated your models.

You should also show the model to the client or others as part of the testing and evaluation process.

Keep Going Until Your Model Meets All of the Design Criteria

1) You might find that you end up <u>changing something</u>, then <u>trying it out</u>, then making <u>another</u> change, and so on. That's just the way it goes sometimes.

2) Here's a summary of the process:

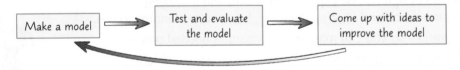

Make a model → Test and evaluate the model → Come up with ideas to improve the model

A very similar process is used to develop a prototype (see p.147) — the difference is that prototypes are testing the <u>real materials</u> and the actual <u>manufacturing processes</u> that are intended to be used to make the final product.

3) Throughout this process, you should be referring to the <u>design specification</u>.

Now You Should Be Ready to Start Making Prototypes

Once you've finished developing your ideas and have a <u>final design</u>, you're <u>ready</u> to start <u>developing</u> a <u>prototype</u> (see p.147). You'll need to have an idea of:

1) The best <u>materials</u>, <u>tools</u> and other <u>equipment</u> to use (and their availability). This might include <u>standard components</u> (see p.64).

2) The <u>assembly process</u> — this is important when it comes to <u>planning production</u> (see p.146).

3) The approximate <u>manufacturing time</u> needed to make each product.

4) How much it should <u>cost</u> to manufacture each product.

Ideas only need to be rough at this stage, as they'll be tweaked quite a lot during prototype development.

Models are repeatedly evaluated against the design specification...

It's worth spending some time getting the models as right as they can be at this stage of the design process, when materials are pretty cheap and easy to work with. It'll save you lots of time and effort in the long run.

Drawing Techniques

To get your design across, you're going to need to draw it on paper. Here are a few techniques to help you communicate your design in the best way possible. Remember, practice makes perfect...

You Can **Develop Design Ideas** with **Freehand Sketches**

1) 'Freehand' means drawing without using any equipment (except a pencil or pen).

2) It's the quickest method of drawing, so it's handy for getting initial design ideas down on paper.

3) You can combine 2D and 3D sketches to explain details.

4) And you can annotate your sketches (add notes) to explain details further, e.g. describing the materials and processes you'd use.

Sketches aren't supposed to be perfect — they only need to get your ideas across.

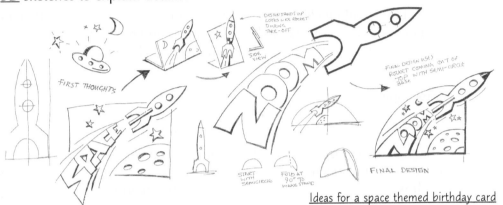

Ideas for a space themed birthday card

Perspective Drawing Uses **Vanishing Points**

1) Perspective drawing tries to show what something actually looks like in 3D — smaller in the distance, larger close up. It does this by using lines that appear to meet at points called vanishing points.

2) These points are in the distance on the horizon line.

One-Point Perspective — for drawing objects head on.

1) Mark one vanishing point.

2) Draw the front view of the object head on.

3) Then draw lines to the vanishing point.

Two-Point Perspective — for drawing objects edge on.

1) Draw a horizon line horizontally across the page.

2) Mark two vanishing points near the ends of the horizon line.

3) Draw the object by starting with the front, vertical edge and then projecting lines to the vanishing points.

4) Remember that vertical lines remain vertical and all horizontal lines go to the vanishing points.

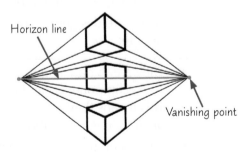

Isometric Drawing Shows Objects at **30°**

1) Isometric drawing can be used to show a 3D picture of an object.

2) It doesn't show perspective (things don't get smaller in the distance), but it's easier to get across the dimensions than in perspective drawing.

3) There are three main rules when drawing in isometric:

- Vertical edges are drawn as vertical lines.
- Horizontal edges are drawn at 30° (from horizontal).
- Parallel edges appear as parallel lines.

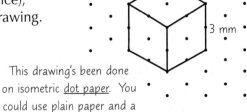

This drawing's been done on isometric dot paper. You could use plain paper and a 30°/60° set square instead.

Drawing Techniques

Some diagrams are drawn specifically for electronic and mechanical systems...

System Diagrams are Flowcharts

1) System diagrams are flowcharts that separate a system into input, process and output boxes. This is useful when developing the basic design for a system.

2) Annotations are used to outline the concept of the system, without going into detail on how exactly you'd go about making it work — this is left to the schematic diagrams (see below).

3) Annotations should include the components or mechanisms that will be used at each stage, and they should describe what the components and mechanisms will do. They should be kept brief and to the point — they don't need to be very detailed.

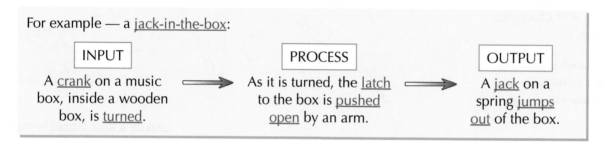

For example — a jack-in-the-box:

INPUT		PROCESS		OUTPUT
A crank on a music box, inside a wooden box, is turned.	⟹	As it is turned, the latch to the box is pushed open by an arm.	⟹	A jack on a spring jumps out of the box.

4) Freehand drawings can be added to system diagrams to help get the point across, but it's really the annotations that are the most important.

Another example of a system diagram is the electronic egg timer on p.30.

Schematic Diagrams Show the Layout of a System

1) Schematic diagrams clearly show the layout of electrical and mechanical systems.

2) A circuit diagram is an example of a basic schematic diagram — it clearly shows how the components are connected up.

3) Repair manuals often use schematics to show the user how the system should be assembled.

4) Schematics are designed to be easy to read and simple to draw. For example:

Schematics are generally drawn later in the design process than system diagrams. For example, you need a basic flowchart outlining an electronic system before you can draw an appropriate circuit diagram for it.

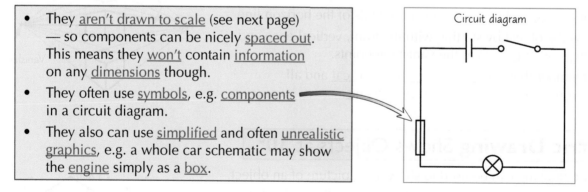

- They aren't drawn to scale (see next page) — so components can be nicely spaced out. This means they won't contain information on any dimensions though.
- They often use symbols, e.g. components in a circuit diagram.
- They also can use simplified and often unrealistic graphics, e.g. a whole car schematic may show the engine simply as a box.

Circuit diagram

 EXAM TIP

Try not to get schematic and system diagrams muddled up...

In the exam, you may be asked to design something. Even if you're not great at drawing, you can still get a lot of the marks by annotating your drawings to explain the details of your design.

More on Drawing Techniques

Here are a few more drawing techniques to help you along. First off, it's drawing things to scale...

Scale Drawings are Used to Draw Big Things (but smaller)

1) To draw a big object on a small piece of paper, you have to scale it down.
2) The object's still drawn in proportion — it's just smaller.
3) The scale is shown as a ratio. For example:

Scale drawings are said to be "drawn to scale".

- A scale of 1:2 means that the drawing is half the size of the real object.
- A scale of 1:4 means that the drawing is a quarter of the size of the real object.
- Anything drawn at 1:1 is full sized.

4) You can also scale things up. A scale of 2:1 means the drawing is twice the size of the real object.
5) The scale needs to be clearly shown on the diagram. It's a ratio, so it doesn't have any units.

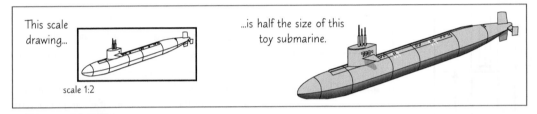

This scale drawing...

scale 1:2

...is half the size of this toy submarine.

Lines on a scale drawing should be labelled with the lengths of the real object — not the lengths of the lines on the paper.

EXAMPLE:

A jet ski is 1.2 m tall. A scale drawing of the jet ski has a height of 40 cm. What is the scale of the drawing?

1) First, convert the measurements to the same units.

Scale drawing height = 40 cm   Jet ski height = 1.2 m = 120 cm

2) Write the measurements as a ratio.

"scale drawing : real object" = 40 : 120

3) Simplify the ratio by dividing each side by the same number.

Both sides will divide by 40.

40 ÷ 40 = 1 and 120 ÷ 40 = 3, so the scale of the drawing is 1 : 3

6) To check you've scaled an object down properly, measure the lengths of the lines in your drawing. If you multiply those lengths by the scale, you should get the dimensions of the real object.

Exploded Diagrams Show How Parts Fit Together

Assembly drawings show how separate parts join together. An exploded diagram is a type of assembly drawing.

This exploded view is also an isometric drawing.

1) Exploded diagrams are always in 3D.
2) You draw the product with each separate part of it moved out as if it's been exploded.
3) Each part of the product is drawn in line with the part it's attached to.
4) Dotted lines show where the part has been exploded from, and therefore where it fits into the overall product.

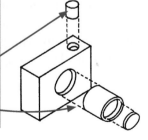

Exploded diagrams are designed so you can use them on their own — you don't need many words to explain how to assemble something. This means that as well as being sent to manufacturers, they're often used for flat-pack furniture instructions.

More on Drawing Techniques

Once you've <u>perfected</u> your <u>design idea</u>, you'll need to produce an <u>accurate working drawing</u> so that the <u>manufacturer</u> can <u>make it</u>. Here's how to wow them (and the examiners) with some <u>super clear</u> drawings...

Orthographic Projection Shows 2D Views of a 3D Object

1) <u>3rd angle projections</u> are used very widely in industry to help the <u>manufacturer</u> understand the design.

2) They show a <u>3D object</u> as a set of <u>2D drawings</u> viewed from <u>different angles</u> — a <u>front</u> view, a <u>plan</u> view (as seen from <u>above</u>) and an <u>end</u> view (as seen from the <u>side</u>).

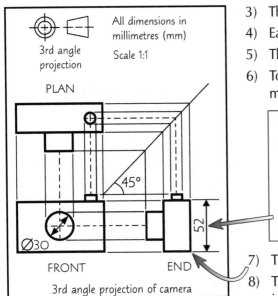

All dimensions in millimetres (mm)

Scale 1:1

3rd angle projection of camera

3) The <u>symbol</u> for <u>3rd angle orthographic</u> projection is: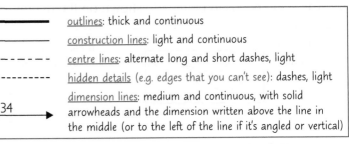

4) Each 2D view is drawn <u>accurately to scale</u>.

5) The <u>dimensions</u> are always given in <u>millimetres</u>.

6) To avoid confusion, lines and dimensions must follow certain <u>conventions</u>:

————————	<u>outlines</u>: thick and continuous
————————	<u>construction lines</u>: light and continuous
— · — · — · — ·	<u>centre lines</u>: alternate long and short dashes, light
– – – – – – – –	<u>hidden details</u> (e.g. edges that you can't see): dashes, light
◄——— 34 ———►	<u>dimension lines</u>: medium and continuous, with solid arrowheads and the dimension written above the line in the middle (or to the left of the line if it's angled or vertical)

7) There's always a <u>gap</u> between the <u>projection lines</u> and the <u>object</u>.

8) The diameter of a <u>circle</u> is shown by the symbol Ø and an arrow inside the circle.

9) To <u>draw</u> a <u>3rd angle orthographic projection</u>, you:

> 1) Draw the <u>front view</u>.
>
> 2) Add <u>construction</u> and <u>centre</u> lines to the <u>right</u> — these help to draw the <u>outlines</u> of the <u>end view</u>, which you should now complete.
>
> 3) Add <u>construction</u> and <u>centre</u> lines going <u>up</u> from the <u>front view</u>.
>
> 4) Draw a line at <u>45°</u> (as shown in the diagram above) in the <u>top right hand corner</u> of the <u>front view</u>.
>
> 5) Draw <u>construction</u> and <u>centre</u> lines going <u>up</u> from the <u>end view</u>. When they <u>reach</u> the <u>45° line</u>, draw them going <u>90°</u> and to the <u>left</u> — these lines and the ones coming up from the front view can be used to draw the <u>plan view</u>.
>
> 6) Now the projections have taken shape, it should be easier to figure out which edges are the <u>hidden details</u> in each of the views — <u>mark these</u> on the drawings (see below).
>
> 7) Add the <u>dimensions</u> — don't forget to use <u>millimetres</u>.

You may be asked to draw the third view when given the others. Draw construction and centre lines from the two views you have (using the 45° line if you need to project between plan and end views), and use these lines to draw the third view.

10) Some objects have <u>hidden details</u> that you can only see from <u>certain views</u>. These can be shown in orthographic projections. For example:

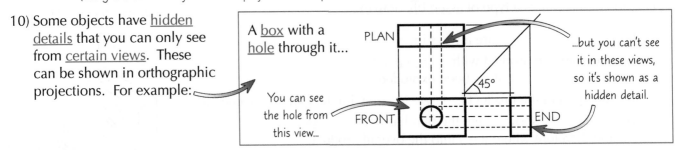

A <u>box</u> with a <u>hole</u> through it...

You can see the hole from this view...

...but you can't see it in these views, so it's shown as a hidden detail.

An orthographic projection is made up of front, plan and end views...

Orthographic projection can take a while to get your head around. It's best tackled with a brew in hand.

Warm-Up and Worked Exam Questions

It's that time again — first, see if you've understood the basics by tackling these warm-up questions.
After that, follow the worked exam questions and see if you can do the questions on the next page yourself.

Warm-Up Questions

1) In design, what is the name for getting stuck on a particular idea?
2) Describe what is meant by evaluating a model.
3) Give one advantage and one disadvantage of using isometric rather than perspective drawing.

Worked Exam Questions

1   A company have produced a scale drawing of a table for a client.

   a) The table will have a width of 1.25 m. The scale drawing of the table has a width of 25 cm.
   Calculate the scale of the drawing. Give your answer as a ratio in its simplest form.

   The measurements need to be in the same units (e.g. cm) before you write them as a ratio.

   Width of the table = 1.25 m = 125 cm

   In the scale drawing this measurement is 25 cm

   Scale of the drawing = 25 : 125 = 1 : 5

   25 : 125 is simplified by dividing both sides by 25.

   [2 marks]

   b) The scale drawing of the table has a height of 18 cm. What will its real height be?

   18 × 5 = 90 cm

   The 1 : 5 ratio can be read as "scale drawing size : real object size", so the height of the table is 5 times larger than its height in the scale drawing.

   [1 mark]

2   A seamstress is designing a one-off dress for a client. She is in the process of making a toile.

   a) Describe how a toile could be used to improve the design of the dress.

   The toile could be put on a mannequin to work out the proportions and fit of the dress.

   Alternatively, the toile could be tried on by the client themselves.

   [2 marks]

   b) i) It's important for the fabric used to make a toile to have similar properties
   (e.g. stretchiness) to the fabric that will be used to make the real garment.
   State **one** other property that a toile fabric should have. Give a reason for your choice.

   It should be cheap, so that the design can be experimented with and improved without it costing too much.

   Toiles should also be lightly coloured, so changes can be marked up onto it easily.

   [2 marks]

   ii) Name a material that is commonly used as a toile fabric.

   Calico

   [1 mark]

Exam Questions

1   Iterative design is a strategy centred around a constant process of evaluation and improvement.

Use sketches and/or notes to give a detailed description of the iterative design process.

[4 marks]

2   The bookcase in **Figure 1** needs to be assembled at home by the consumer.
Instructions are needed to help the consumer assemble it.

a)   Name the style of drawing that would be most suitable for the
instructions. Explain why this style of drawing is suitable.

Name: ....

Explanation: ....

....

[2 marks]

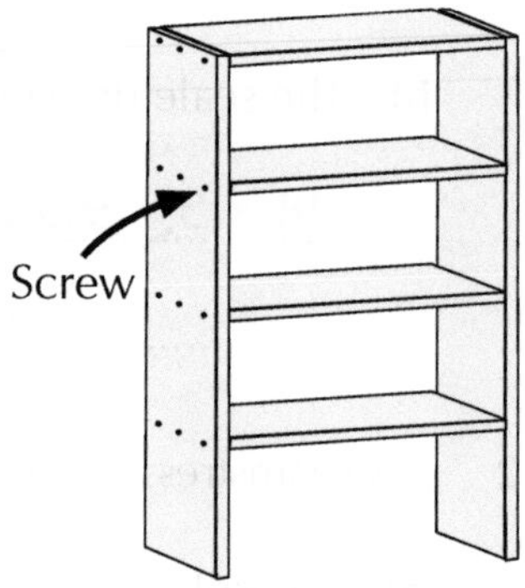

Screw

Figure 1

b)   Draw assembly instructions for the bookcase using this style of drawing.

[3 marks]

Manufacturing Specification

A picture might be worth a thousand words, but working drawings don't give the manufacturer everything they need. The manufacturing specification provides the extra information required to make the product...

There are Loads of Things that Manufacturers Need to Know

A manufacturing specification can be a series of written statements, or working drawings and sequence diagrams (see next page). It has to explain exactly how to make the product, and should include:

1) clear construction details explaining exactly how to make each part,
2) materials — which materials to use for each part and how much will be needed,
3) equipment — what's needed at each stage,
4) sizes — precise dimensions of each part in millimetres,
5) tolerances (see p.49) — the maximum and minimum sizes each part should be,
6) finishing details — any special information, such as 'laminate the paper with aluminium',
7) quality control instructions (see p.49) — what needs to be checked, and how and when to check it,
8) costings — how much each part costs, and details of any other costs involved.

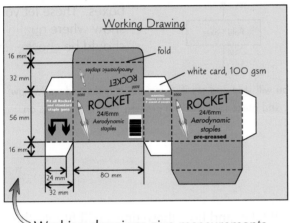

Working drawings give measurements, materials and construction details (e.g. where to make folds).

Spreadsheets are great for working out costings.

Rocket staple box [600 ___] cost

Item	Cost (£) per 1000	Cost (£) per 5000
Raw Materials		
Laminated grey board	6.66	33.3
Red ink	8.08	40.4
Blue ink	2.02	10.1
Black ink	2.02	10.1
Total	18.78	93.9
Fixed costs		
Printing plate	42.42	42.42
Cutting tool	13.13	13.13

Each Stage Needs to be Planned in Detail

Take each stage of the manufacturing process and plan it in detail. You need to think about:

1) how long each stage will take,
2) what needs to be prepared before you can start each stage,
3) how you will ensure consistency and quality, e.g. using jigs, formers and measuring tools,
4) how you will do quality control checks,
5) what health and safety precautions you will have to take to be safe when making your product.

Remember, your methods would probably change if you were going to produce your design in quantity.

The manufacturing specification should be as detailed as possible...

As nice as working drawings may be, you can't just give a manufacturer a picture and expect them to bring the design to life. Make sure you know everything that should be included in the manufacturing specification.

Manufacturing Specification

Making a <u>few</u> examples of your product is (relatively) <u>easy</u>. But <u>mass-producing</u> it is much harder and takes a lot of careful <u>planning</u>. This is where <u>sequence diagrams</u> such as <u>flowcharts</u> and <u>Gantt charts</u> come in handy...

Charts Help to Plan the Manufacturing Process

Work Order

1) A <u>table</u> or <u>flowchart</u> can be used to plan the <u>work order</u> — the <u>sequence</u> in which tasks should be carried out.

2) They can also include <u>tools</u>, <u>quality control</u> stages, <u>safety</u>, and so on.

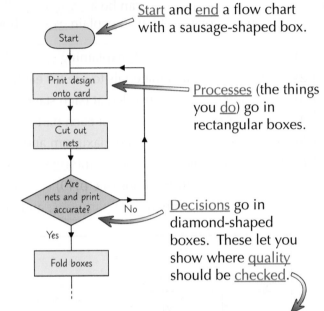

<u>Start</u> and <u>end</u> a flow chart with a sausage-shaped box.

<u>Processes</u> (the things you <u>do</u>) go in rectangular boxes.

<u>Decisions</u> go in diamond-shaped boxes. These let you show where <u>quality</u> should be <u>checked</u>.

Prototype staple box			
<u>Day</u>	<u>Process</u>	<u>Tools needed</u>	<u>Quality ch</u>
1	Print designs	Airbrush, pens, dry transfer lettering	Make s
2	Cut out net	Scalpel, metal rule	Chec
	Score and fold net	Scissors, metal rule	

The diamond-shaped boxes show where you will stop and see if your product looks and works how it should. If you find it doesn't, go back and make sure it's done properly before you move on.

Gantt Charts

1) You also need to work out <u>how long</u> each stage will take, and how these times will fit into the <u>total time</u> you've allowed for production. One way to do this is with a <u>Gantt chart</u>.

2) The <u>tasks</u> are listed down the <u>left-hand</u> side, and the <u>timing</u> is plotted across the <u>top</u>.

3) The <u>coloured squares</u> show <u>how long</u> each task takes and the <u>order</u> they're done in. In this chart, each square represents 5 minutes.

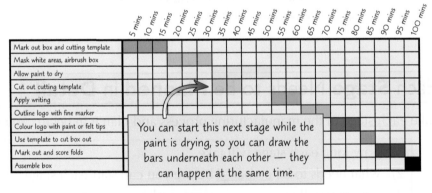

You can start this next stage while the paint is drying, so you can draw the bars underneath each other — they can happen at the same time.

Make sure you know how to read the timings on a Gantt chart — the times at the top show how much time has passed in the overall production process — not how long each stage will take. For example, 55 minutes into production, the writing should be being applied, and this should take about 10 (not 55-60) minutes to complete.

Gantt charts let you see all the stages in a process...

REVISION TIP

A really good way to revise Gantt charts is to have a go at making one. It could be for anything — revision, household chores, homework etc. It might just end up being useful, too.

Developing Prototypes

Next up, it's <u>developing prototypes</u>. I know it sounds a bit dull, but this is <u>super useful</u> for your <u>non-exam assessment</u> (<u>NEA</u>, see p.156), where you'll be developing a prototype of your own, so it's <u>well worth a read</u>...

Prototypes are Full-Size, Working Products or Systems

1) Prototypes are a <u>step further</u> than the models on page 137 — they're <u>full-size, fully-functioning products</u> or <u>systems</u>.

2) They're made using the <u>materials</u> and <u>manufacturing methods</u> that are intended for the <u>final product</u>.

3) This allows you to <u>test</u> the <u>product</u> and the <u>production methods</u> you're going to use to make sure they're as you want them.

The final product you make in your project is a prototype of your design.

Prototypes Allow You to Evaluate the Manufacturing Process

1) Making a prototype allows you to <u>check</u> that the <u>manufacturing specification</u> (see p.145) is <u>correct</u>.

2) Any <u>problems</u> identified during this process will need to be <u>solved</u>, and the manufacturing specification <u>modified</u> to include the changes.

3) Making a prototype can also help you calculate some of the <u>manufacturing costs</u> — this is important so you <u>don't</u> end up with a <u>final product</u> that costs <u>far too much</u> to make. In industry, these costs can include:

- <u>Materials</u> and <u>components</u>
- <u>Labour</u>
- <u>Packaging</u>
- <u>Cost</u> of new <u>equipment</u>
- <u>Energy</u>
- <u>Waste disposal</u>

The design specification should contain information on the manufacturing costs.

You Can Evaluate Your Design Using Your Prototype

1) <u>Prototypes</u> are <u>evaluated</u> by checking that they meet the criteria set out in the <u>design specification</u> (see p.128). It's best to go through these criteria <u>one by one</u>.

2) You'll probably find that parts of your design <u>don't work out</u> the way you wanted them, and so don't fit the <u>design specification</u>. If so, you need to <u>think</u> about the <u>improvements</u> you could make. You need to be able to <u>reject</u> the part of the design that <u>doesn't work</u> and <u>justify</u> an <u>alternative</u>. For example:

Remember, the design specification contains the conditions a product should meet, e.g. the product should be functional, safe to use, have good aesthetics, etc.

Design Specification for a Reusable Shopping Bag
- The finished bag must be retailed for £3 or less.
- The bag must be made out of environmentally-friendly material.
- It must be able to hold 4 kg of stuff.

Maybe the <u>plain seams</u> (see p.102) <u>attaching</u> the <u>handles</u> to the bag <u>aren't strong enough</u> to hold the 4 kg weight. Instead, you could use a <u>flat-felled seam</u> to attach the handles to the bag, as this is a <u>stronger</u> and <u>more durable</u> method of <u>attachment</u>.

3) You also need to make sure that the prototype <u>meets</u> the requirements of the <u>design brief</u> — what the <u>client wants</u> or <u>needs</u> (see p.127).

In your project, you'll be asked to <u>evaluate</u> your prototype, so this stuff should come in handy there too...

Developing Prototypes

You Should Also Get **Feedback** from **Other People**

1) Prototypes are useful as they allow <u>other people</u> to <u>try out</u> your product.

2) <u>Feedback</u> from the <u>client</u>, <u>potential customers</u> and <u>experts</u> in the industry (see p.136) provide <u>additional suggestions</u> on how you could <u>modify</u> your design.

3) This will hopefully <u>improve</u> the product so that it's <u>marketable</u> — <u>appealing</u> to the target market, is <u>fit for purpose</u> and will <u>sell</u> (hopefully).

> *You should use questionnaires and interviews to get feedback and record your findings in a table so it's easy to read.*

- E.g. potential customers might test the reusable shopping bag prototype and say that the <u>cord handles</u> are <u>good</u> for carrying a <u>load of shopping</u>, but when the bag is <u>empty</u>, the cords make it <u>bulky</u> and <u>difficult to fold away</u>.
- This means you can <u>justify</u> making the handles out of <u>flexible cotton straps</u> instead, although a prototype with these should be <u>tested</u> and <u>evaluated</u> to make sure there aren't other problems with these.

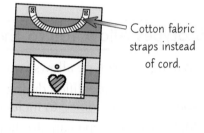

Cotton fabric straps instead of cord.

Keep Going Until You Get it **Just Right**

1) You might find that you end up <u>modifying</u> something, then <u>evaluating</u> it, then making <u>another modification</u> and <u>testing</u> that, and so on. This <u>repetitive</u> process is known as <u>iterative design</u> (see p.135) and is all part of <u>developing</u> a good product.

2) Depending on your <u>time</u> and <u>resources</u>, you could make any number of <u>alternative</u> prototypes. Once you've <u>evaluated</u> them <u>all</u>, go back and select the <u>best</u> one.

3) When you're doing this for real with your own product, you need to keep a <u>record</u> of <u>what you find out</u> from each prototype and the <u>changes</u> you make to your design as a result. Include your <u>justification</u> too, e.g. <u>why</u> each new prototype <u>better fits</u> the <u>specifications</u>.

> *You could even create a database of photos of each prototype to help you remember what you did at each stage.*

4) Once a product is as <u>perfect</u> as it can be, it's time to consider <u>production</u> on a <u>larger scale</u>.

5) Testing prototypes is very important in <u>industry</u>, as any <u>mistakes</u> that aren't picked up will <u>cost a lot</u> of money to <u>put right</u>.

Designing a new product can involve lots of prototypes...

Developing a really good prototype involves looking at each prototype you make critically and making improvements. Sometimes, you can take one step forward and two steps back if your changes mean a different point on the design specification isn't met, so think about any knock-on effects a change might have.

Using Materials Efficiently

Efficient use of a piece of material means making as many parts or products as possible with it...

Wasting Materials Means Losing Money

1) Any material that manufacturers waste means a loss of money — so when they're making products, they have to make the best use of the materials.

2) If manufacturers can make their product for less money, it means they can make more profit or afford to sell it to us more cheaply.

3) Careful planning of where to cut (nesting — see next page) and accurate marking out can reduce waste.

It's important to reduce waste for environmental reasons too (see p.6).

There are Loads of Different Tools for Marking Out

1) Marking out is making a mark so you know where to cut, drill or assemble your material.

2) This helps to make sure that products or parts are made accurately, consistently and to a high quality.

3) Marking out may be as basic as using a pencil on a piece of wood (you shouldn't use a pen as it seeps into the wood and you can't get it out) or a felt tip on some plastic — but there are lots of other ways:

SCRIBER
— used like a pencil but it scratches a mark into metal and plastic.

ENGINEER'S BLUE and MARKING BLUE are dyes you put on metal so any scribe marks show up better.

PATTERN MASTER
— used to help you draw paper patterns. The curved edges are for marking out smooth curves and the parallel lines are for marking out extra width for seam allowance.

SLIDING BEVEL
— can be set at any angle to guide a marking knife.

ODD-LEG CALIPER
— marks a line parallel to an edge.

TAILOR'S CHALK
— used to transfer markings onto a fabric that you can remove later.

MARKING KNIFE
— used to score lines in wood (it cuts the fibres to stop them splitting during sawing).

MARKING GAUGE
— scratches a line in wood parallel to an edge.

Things need to be cut or shaped to the correct size — so measure carefully when you're marking out. Steel rules can be used to measure lengths. Combination squares can be used to check things are level, measure angles, and find the centre of circles.

TRY SQUARE
— helps to accurately mark out right angles (90°).

TEMPLATES and PATTERNS (see p.53-54) — templates can be drawn round to mark out the same shape. Patterns can be cut round.

4) You can also mark out with the help of reference points, lines and surfaces, and coordinates (see p.53). You don't always need to mark out though — jigs (see p.54) can be used to guide tools instead.

When Marking Out, Remember...

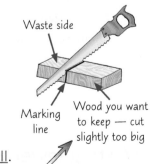

Waste side

Marking line

Wood you want to keep — cut slightly too big

1) Always measure things out twice — that way, you're less likely to make a mistake.

2) Keep the marks as thin as possible — otherwise your cutting might be inaccurate.

3) Remember to get rid of the marks when you've finished so it doesn't look untidy.

4) If possible, mark out on a surface that the user won't see — for example on the back of a door, or the inside surface of a box.

5) When you cut, always do it on the waste side of the line so it doesn't end up too small. If it's a bit big, you can always file it down to size — but if it's too small you'll have to cut it again.

6) And remember, you can't mark and cut everything perfectly — nobody's perfect. But it's important to work within the required tolerance (see p.49).

Using Materials Efficiently

Arranging Things Efficiently Helps to Reduce Waste

1) When <u>batch</u> or <u>mass producing</u> a <u>product</u> (see p.47 and 48), the <u>same</u> shapes are <u>cut</u> out <u>again and again</u> often from stock forms.

2) Waste can be reduced by making as <u>many products</u> as possible from the <u>least amount of material</u>. For example, manufacturers can try to cut as many shapes from <u>one sheet</u> of material as possible.

3) Some shapes will <u>tessellate</u> — this means that <u>repeats</u> of the shape <u>fit together</u> without any <u>gaps</u> or <u>overlapping</u> pieces, e.g. a hexagon. These shapes can be cut from sheets of material with <u>very little waste</u>.

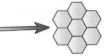

4) When <u>planning</u> the cutting and shaping of a material, manufacturers try to come up with the most <u>efficient arrangement</u> of shapes to <u>minimise waste</u> — this process is called <u>nesting</u>.

5) To <u>work out</u> which <u>arrangement</u> is <u>best</u>, you can work out the <u>area</u> (if it's a <u>sheet</u>) or <u>volume</u> of <u>material</u> that is <u>going to waste</u>:

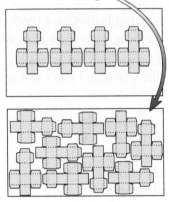

- First, calculate the <u>area</u> (or <u>volume</u>) of <u>one shape</u>.
- <u>Multiply</u> this by the <u>number of shapes</u> you can <u>fit</u> onto the <u>material</u> — this gives the amount of material that <u>can be used</u>.
- Work out the <u>area</u> (or <u>volume</u>) of <u>material</u> you <u>started with</u>.
- Then <u>subtract</u> the amount of <u>useful material</u> from this — this leaves the amount of <u>waste material</u>.

The arrangement with the <u>smallest amount of waste</u> should be <u>chosen</u>.

Don't worry — you're allowed to use a calculator in the exam.

You can fit a fair few cube nets onto this sheet of material if you arrange them right.

EXAMPLE: **The triangle on the right can fit onto a 21 cm × 29.7 cm sheet of paper, 23 times. Calculate the amount of paper wasted in a single sheet.**

Area of a triangle = ½ × width × height = ½ × 6 × 7.5 = 22.5 cm²
Total area of triangles = 22.5 × 23 = 517.5 cm²
Area of paper = length × width = 21 × 29.7 = 623.7 cm²
So the amount of paper wasted = 623.7 - 517.5 = 106.2 cm²

Remember to write your units, e.g. cm².

7.5 cm

6 cm

---

6) One way to <u>quickly work out</u> an <u>efficient layout</u> is to use <u>CAD</u> — e.g. CAD can be used in <u>lay planning</u> to find an efficient way of laying out <u>pattern pieces</u> onto a <u>sheet of fabric</u>.

7) Remember to include any <u>extra material</u> needed to <u>join</u> materials together when <u>marking out</u> your shapes.

- For example, <u>nets</u> used to make packaging (see p.67) have extra <u>flaps</u> of material that are <u>glued</u> to <u>another part</u> of the <u>packaging</u>. This <u>holds it</u> all <u>together</u>.
- In clothes, extra fabric is needed for the <u>seam allowance</u> (see p.102).
- In <u>joints</u> (e.g. in wood or metal), extra material is given so there can be an <u>overlap</u>.

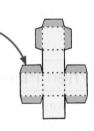

Wasting fewer materials is good for the environment...

Marking out is pretty important. It's something you need to do accurately when hand-making prototypes. You'll end up wasting less time and materials, as you're more likely to get it right first time.

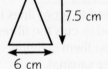

Working Safely

You need to <u>think</u> about <u>safety</u> when using different <u>tools</u> and <u>equipment</u>.
This is <u>important</u> for your <u>own safety</u> but also the safety of the <u>people around you</u>.

Safety is **Essential** when you **Make** a Product

A lot of this is common sense, but it's <u>incredibly important</u> stuff so pay attention...

Wear Protective Clothing and Equipment

1) With <u>hazardous materials</u> wear a <u>face mask</u> or <u>goggles</u> and <u>protective gloves</u>. These will protect the <u>skin</u> and <u>eyes</u>, e.g. from <u>splashes</u> and <u>spillages</u>.

2) If the material is <u>hot</u>, wear <u>protective gloves</u> and an <u>apron</u>. For some jobs, e.g. welding, you should also wear a <u>tinted face shield</u>.

3) When using cutting machines, wear <u>chainmail gloves</u> to protect your hands.

4) When using <u>pins</u> and <u>needles</u>, use a <u>thimble</u>.

5) When using <u>noisy machinery</u>, wear <u>ear protection</u>.

Be Careful with Tools and Machinery

1) When working with tools and machinery, make sure you have your <u>sleeves rolled back</u>, <u>ties tucked in</u> or taken off, <u>apron strings tucked in</u>, <u>necklaces</u>, <u>watches</u> and <u>rings taken off</u>, and <u>long hair tied back</u>.

2) Never leave any machines <u>unattended</u> while switched on.

3) Never use a machine or hand tool unless you've been <u>taught</u> how to.

4) Know how to <u>switch off</u> machines in an <u>emergency</u>.

5) Don't <u>change parts</u> (e.g. drill bits) on a machine until you've <u>isolated</u> it from the mains.

6) Always <u>secure</u> work safely — e.g. clamp work securely before cutting.

7) Use <u>guards</u> on <u>machines</u> that have them, e.g. sanding machines and pillar drills.

8) Use a <u>dust extractor</u> if the process produces dust. (It might be sensible to wear a <u>dust mask</u> and <u>eye protection</u> too.)

9) <u>Carry</u> and <u>store</u> sharp tools safely.

Handle Materials and Waste Sensibly

1) <u>Choose</u> your materials sensibly — only use <u>hazardous materials</u> when absolutely <u>necessary</u>.

2) Make sure materials are <u>safe to handle</u>, e.g. <u>file down</u> rough edges before starting work.

3) When <u>storing</u> material, make sure it's <u>put away safely</u> so it can't fall and injure anyone. Keep <u>flammable liquids</u> away from <u>naked flames</u> and red-hot heating elements.

4) When using <u>toxic chemicals</u> such as dyes, finishes and solvents, you need adequate <u>ventilation</u> to avoid <u>inhalation</u> of vapours.

5) Dispose of <u>waste</u> properly so that it doesn't <u>harm</u> the <u>environment</u>.

CAUTION

FLAMMABLE LIQUIDS

Most chemicals will have guidelines and regulations telling you why they're hazardous and how to use, store and dispose of them safely.

All making activities involve some sort of hazard...

Making a product is full of risks and hazards. Some can be avoided by being careful and sensible. But you must pay attention to labels or safety instructions on chemicals and equipment — they're there for a reason.

Working Safely

Risk Assessments are also Important

1) A risk assessment is used to identify and minimise any risks when working.

2) When writing a risk assessment, think about:

- What could be a hazard?
- What precautions could be taken to make sure the risk is minimised?

3) Here's part of a risk assessment for making a carved wooden box.

4) Risk assessments are carried out:

Risk assessment for project

Hazard	Precaution — how to reduce the risk
Clothing could get caught in the sanding machine.	Tuck clothes in and wear an apron.
Fine dust created when using a sanding machine.	Wear a mask and use a dust extractor.
Fingers could be cut when using a craft knife.	Use a safety rule to protect fingers.

- On the product during the design process, to assess potential risks to the end user.
- By manufacturers, for the materials, chemicals and equipment that workers will have to use during the manufacture of the product.

You Can Carry Out Tests to Ensure Safety

Even in school, there are things you can do to make sure the product you're designing is safe.

1) Research your materials carefully and test them to make sure they're suitable, e.g. see how far they'll bend, check whether they catch fire easily, etc.

2) Use standard components (see p.64) wherever you can, because these have already been tested by the manufacturer — this helps make sure that safety standards are met.

3) Make prototypes and carry out real-life simulations, e.g. if you're making a cot, check that a toddler's finger can't get trapped in the gaps between the bars (using a metal rod of about the same width as a toddler's finger, for example).

4) Get electrical items PAT tested (PAT stands for Portable Appliance Test). The test makes sure that portable electric products won't hurt anyone if used properly. School technicians can sometimes do this.

5) You can test products using CAD software. If you've selected the right materials, you can simulate stress tests, such as how well a product will stand up to an impact.

Risk assessments help you to manufacture the product safely...

Some of this stuff is common sense, but that doesn't mean it isn't important. You should know what a risk assessment is and how to write one. Come up with some safety top tips for working in a tech workshop. Then write a risk assessment for a particular activity you may do here (e.g. welding metal) using these tips along with any other relevant safety precautions you can think of. Remember to say what the hazard is before writing the precaution you'd take.

Warm-Up and Worked Exam Questions

Congratulations — you've just covered the last bits of content in the entire book. Good work.
Time to take a deep breath and give these warm-up and exam questions everything you've got...

Warm-Up Questions

1) What are prototypes?
2) Explain what is meant by nesting.
3) Explain how standard components can be useful in making sure your product is safe for use.

Worked Exam Questions

1   Marking out is one way of reducing the waste produced when making a product.

Explain how marking out can help to reduce waste.

Marking out helps to make sure products are cut accurately and within the

required tolerance. This reduces waste as incorrectly sized products

would have to be thrown away.

[2 marks]

2   **Figure 1** shows the manufacturing process for making a storage box.

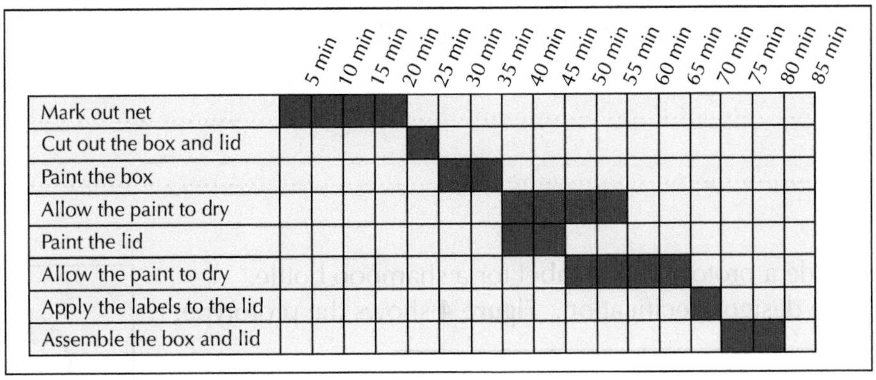

	5 min	10 min	15 min	20 min	25 min	30 min	35 min	40 min	45 min	50 min	55 min	60 min	65 min	70 min	75 min	80 min	85 min
Mark out net	■	■	■														
Cut out the box and lid				■													
Paint the box					■	■											
Allow the paint to dry							■	■									
Paint the lid									■								
Allow the paint to dry										■	■	■					
Apply the labels to the lid													■				
Assemble the box and lid														■	■		

Figure 1

a)   Name the type of sequence diagram shown in **Figure 1**.

A Gantt chart

[1 mark]

b)   Using this sequence diagram, state how long it should take to paint the box.

10 minutes

The row of the chart that involves painting the box has two shaded squares.
Each square is worth 5 minutes, so it should take 10 minutes to complete.

[1 mark]

c)   Suggest **one** change that could be made to the manufacturing process to reduce
the total time it takes. Explain your reasoning.

The lid could be painted before the box is painted, to allow the labels to be applied

while the box is drying.

To make the manufacturing process quicker, the order in which the tasks
are completed needs to change so that there is more overlap between tasks.

[1 mark]

Exam Questions

1   Triangles are to be cut from the sheet of material shown in **Figure 2**. The triangles
     need to be arranged so that as many of them as possible can be cut from the sheet.

   a)   Repeat the triangle shown in **Figure 2** to show the arrangement
        that results in the least possible waste material.

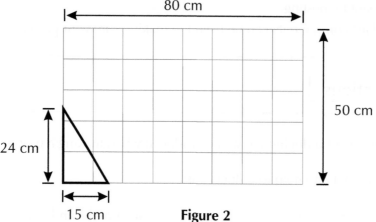

Figure 2

[1 mark]

   b)   Calculate the area of **one** triangle.

        ....

        ....

[1 mark]

   c)   Using your answers to **a)** and **b)**, calculate the minimum amount of material wasted
        from cutting the triangles from the sheet in **Figure 2**.

        ....

        ....

[3 marks]

2   A designer has made a prototype of a label for a shampoo bottle.
     Figure 3 shows the design specification. **Figure 4** shows the prototype.

The label should:
*   show the name of the product and what it smells of
*   have a futuristic or scientific appearance
*   include an image of glossy hair

Figure 3

Figure 4

Evaluate the prototype of the label against the design specification.

Remember, 'evaluating' a prototype against a specification means checking whether each point on the specification is met by the prototype.

....

....

....

....

....

[4 marks]

Revision Questions for Section Eight

Phew — <u>Designing and Making</u> was a mammoth of a section. Here's some revision questions, just for fun...
* Try these questions and <u>tick off each one</u> when you <u>get it right</u>.
* When you've done <u>all the questions</u> for a topic and are <u>completely happy</u> with it, tick off the topic.

Designers, User Needs, Design Briefs and Specifications (p.123-128) ☐

1) Give an example of a designer who designed household items.
2) Suggest how a can opener could be made suitable for infirm elderly people.
3) What information should a design brief include?
4) Give one example of a type of market research.
5) List two things you can find out about a product during product analysis.

Market Research and Product Analysis (p.129-132) ☐

6) Describe a potential drawback of asking open questions on a questionnaire.
7) Give three things that should be considered when analysing a product.
8) What is the term for assembling parts separately and then adding them to the product later?

Design Strategies and Developing a Design Idea (p.135-138) ☐

9) Which design strategy focuses on breaking down the design process into several different stages?
   a) Iterative design   b) User-centred design   c) Systems approach
10) Iterative design is a circular process. Explain what is meant by this.
11) Give one reason why it's important to collaborate with the client early on in the design process.
12) a) What is meant by modelling?   b) Why is modelling useful when developing a design?
13) Once a final design has been developed, what is the next stage in the design process?

Drawing Techniques (p.139-142) ☐

14) What is freehand drawing?
15) What do perspective drawings try to show?
16) Give one feature of schematic diagrams that makes them easy to read and understand.
17) The dimensions of a wooden box are 60 × 100 × 20 mm. Jo is doing a scale drawing of the box using a scale of 1:4. What should the dimensions of the drawing be?

Manufacturing Specifications, Prototypes, Using Materials Efficiently and Safety (p.145-152) ☐

18) a) What is a manufacturing specification?
   b) Give three things that a manufacturing specification should include.
19) On a work order flowchart, how would you show where quality control should take place?
20) Give two reasons why it's useful to make prototypes using the manufacturing specification.
21) a) What is meant by marking out?
   b) Suggest a piece of equipment that you could use to mark out on wood.
22) Explain what it means if a shape tessellates.
23) Suggest a piece of protective clothing or equipment that you would use when:
   a) handling hot materials   b) using cutting machinery
24) a) What two things should be written down in a risk assessment?
   b) Suggest why risk assessments are carried out on the product during the design process.

Assessment Advice

Unlike most subjects, in Design and Technology you actually get to <u>design</u> and <u>make</u> something <u>useful</u>.

The **Non-Exam Assessment** is Worth **50%** of your **GCSE**

1) The '<u>non-exam assessment</u>' (or <u>NEA</u> for short) is a <u>design and make</u> task
— this involves making a <u>working prototype</u> and a <u>design portfolio</u>.

2) At the end of the first year of your course, you'll be given a choice of '<u>contextual challenges</u>'. These are <u>general topic areas</u>, e.g. 'in the workplace'. You'll <u>choose one</u> of these to <u>investigate</u> in your NEA.

3) Your teacher will give you as much help as they're allowed to by the exam board, so do <u>ask them</u>...
but mostly it's <u>up to you</u> to make a <u>good job</u> of your NEA. You can dip into this book for extra help.

4) <u>Section Eight</u> is focused on the <u>design and make</u> stuff, so it's a good place to <u>start</u>.

5) The <u>rest</u> of the book contains <u>useful information</u> for your NEA too though.
If you're wondering about a particular <u>detail</u>, e.g. say, the properties of different woods,
it's probably quickest to look that up in the <u>index</u> and go straight to those pages.

Only Put **Relevant Stuff** in your **Portfolio**

1) Your <u>teacher</u> will give you plenty of <u>guidance</u> about what to put into your <u>portfolio</u>.
This is <u>documentation</u> of your work, which can be either <u>printed</u> or <u>digital</u>.

2) Your portfolio should lay out <u>each stage</u> of the <u>designing</u> and <u>making</u> of your prototype,
including any <u>decisions</u> that you make as you go along. For example, it should include:

- A concise <u>summary</u> of <u>research</u> into your <u>contextual challenge</u>,
including the <u>needs</u> and <u>wants</u> of the intended users, and a <u>design brief</u>.
- A <u>design specification</u> (see p.128), including how your <u>research</u> findings <u>helped</u> you to <u>decide</u> on the <u>design criteria</u>.
- <u>Documentation</u> (including photos) of any <u>models</u> you make. Don't just include the ones that worked
— the ones that <u>didn't quite work</u> are useful, as you can explain <u>what was wrong</u> and how you fixed it.
- A <u>manufacturing specification</u> (see p.145) that explains the different stages of
the making process to show <u>how you constructed</u> your prototype.
- An <u>evaluation</u> of your prototype, including <u>testing</u> it against what you originally set out to create in your <u>design brief</u>.

3) It's recommended that your portfolio is <u>20 pages</u> of <u>A3</u> (or the equivalent in A4)
— you'll need to be <u>thorough</u> but write <u>concisely</u> to stick to this.

4) <u>Check</u> that you've used the right <u>technical words</u> and <u>spelled</u> things correctly.
And make sure you've <u>explained things clearly</u> — get someone who
<u>knows nothing</u> about your project to read it and see if it <u>makes sense</u>.

You should always back up the decisions you make with evidence — e.g. a photo of a model that might not function properly, market research, etc.

The **Exam** is Worth **50%**

1) In the exam you could be tested on <u>anything</u> you've learned during the course — materials, tools,
how to design things, how to make things, health and safety, environmental issues...

2) This book can help you to <u>learn</u> this stuff — and there are <u>questions</u> for you to <u>check</u> you've taken it in.

3) There are <u>answers</u> to these questions at the back of the book. There's a <u>glossary</u> at the back too, in case
you need to sort out your mass production from your continuous production.

4) The <u>exam technique</u> section (pages 157-158) has some <u>worked examples</u> of exam-style questions,
and some hints on how to make sure you get <u>top marks</u>.

Non-Exam Assessment — it's all your assessment outside the exam...

In your NEA, 60% of the marks are for designing and making your prototype. The rest of the marks are for investigating design ideas, your design brief and specification, and the analysis and evaluation of your work.

Exam Technique

Here's What to **Expect** in the **Exam**

1) At the end of your GCSE Design and Technology course you'll have to sit an exam.
   It's worth 50% of your total mark. The exam is worth 100 marks and is 2 hours long.

2) The questions in the exam will test you on all areas of the course. The exam is split into three sections
   which test you on different parts of the course. Each section contains different types of questions:

Section A

- Consists of a mixture of multiple choice questions, worth one mark each, and short answer questions.
- The multiple choice questions have one correct answer and three incorrect answers.
  Only pick one answer — if you pick more than one you won't get the mark.
- 20 marks are available in total.

Section B

- Consists of a few short answer questions and one
  extended response question, which is worth more marks.
- Some questions may offer a choice on what to write about —
  e.g. a choice of materials, products, etc. It's worth spending a
  moment reading the whole question to figure out which option
  to choose — pick the one you could write the best answer for.
- 30 marks are available in total.

Section C

- Consists of a mixture
  of questions including
  short answers and
  extended response
  questions.
- 50 marks are
  available in total.

Remember These **Tips**

1) Always read every question carefully.
   Don't write an essay about modelling if it's asking about developing prototypes.

2) Use the number of marks as a guide on how much to write.
   Try to write a point for each mark. For example, if the question's worth 3 marks, write 3 points — not just a single sentence.

3) Write your answers clearly, using good grammar, spelling and punctuation.
   If the examiner can't read your answer, you won't get any marks, even if it's right.

4) Use the correct terminology.
   Know your technical terms — CMYK, MDF, PIC... you can't write about these if you don't know what they mean.

5) Make your sketches clear.
   Draw clearly and label your sketches — it makes it easier for the examiner to see what you're trying to get across.

6) Don't panic.
   If you really can't do a question, just leave it and move on to the next one. You can come back to it at the end.

Understand the **Command Words**

Questions will often use command words — these words tell you how to answer the question. If you don't
know what they mean, you might not answer the question properly. Here are some you might come across:

State You should give a short answer or list
— you don't need to explain why.

Define You should give a clear, precise meaning
of the word or phrase.

Outline You should give a brief summary of a process.

Explain You should give reasons to show why.

Describe You should give a detailed description of something.

Discuss You should make a balanced argument
covering a range of opinions.

Assess / Evaluate You should use evidence and your own
knowledge to come to a conclusion.

Exam Technique

Plan Out Your Answer to Extended Response Questions

Extended response questions are longer questions worth 6 or more marks and with a scary number of dotted lines underneath... They often use command words like discuss, evaluate or assess (see previous page).

1) Your answer must be well-written (good spelling, grammar and punctuation) and well-structured.
2) Before you start an extended response question, jot down the points you want to make and plan your answer to help structure it well and avoid repeating things.
3) You might have to weigh up the advantages and disadvantages of something, or cover both sides of an argument then form your own opinion.
4) Make absolutely sure you're answering the question and not just waffling on.

Here's an example:

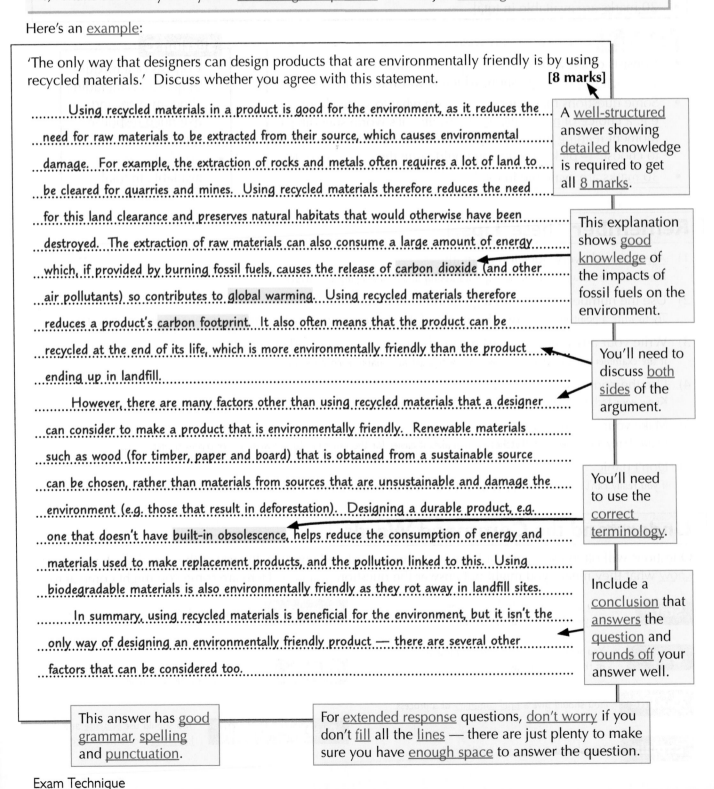

'The only way that designers can design products that are environmentally friendly is by using recycled materials.' Discuss whether you agree with this statement. **[8 marks]**

Using recycled materials in a product is good for the environment, as it reduces the need for raw materials to be extracted from their source, which causes environmental damage. For example, the extraction of rocks and metals often requires a lot of land to be cleared for quarries and mines. Using recycled materials therefore reduces the need for this land clearance and preserves natural habitats that would otherwise have been destroyed. The extraction of raw materials can also consume a large amount of energy which, if provided by burning fossil fuels, causes the release of carbon dioxide (and other air pollutants) so contributes to global warming. Using recycled materials therefore reduces a product's carbon footprint. It also often means that the product can be recycled at the end of its life, which is more environmentally friendly than the product ending up in landfill.

However, there are many factors other than using recycled materials that a designer can consider to make a product that is environmentally friendly. Renewable materials such as wood (for timber, paper and board) that is obtained from a sustainable source can be chosen, rather than materials from sources that are unsustainable and damage the environment (e.g. those that result in deforestation). Designing a durable product, e.g. one that doesn't have built-in obsolescence, helps reduce the consumption of energy and materials used to make replacement products, and the pollution linked to this. Using biodegradable materials is also environmentally friendly as they rot away in landfill sites.

In summary, using recycled materials is beneficial for the environment, but it isn't the only way of designing an environmentally friendly product — there are several other factors that can be considered too.

A well-structured answer showing detailed knowledge is required to get all 8 marks.

This explanation shows good knowledge of the impacts of fossil fuels on the environment.

You'll need to discuss both sides of the argument.

You'll need to use the correct terminology.

Include a conclusion that answers the question and rounds off your answer well.

This answer has good grammar, spelling and punctuation.

For extended response questions, don't worry if you don't fill all the lines — there are just plenty to make sure you have enough space to answer the question.

Practice Paper

Once you've been through all the questions in this book, you should feel pretty confident about the exam. As final preparation, here is a **practice paper** to give you a taste of what the exam will be like. Your exam might not look exactly like this paper, but it will give you some great practice with all of the topic areas.

GCSE Design and Technology

In addition to this paper, you should have:
• Normal writing equipment
• A calculator
• A ruler

Centre name					
Centre number					
Candidate number					

Time allowed:
• 2 hours

Surname	
Other names	
Candidate signature	

Instructions to candidates
• Write your name and other details in the spaces provided above.
• Answer **all** questions in the spaces provided.
• Do all rough work on the paper. Cross through any rough work that you do not want to be marked.

Information for candidates
• The marks available for each question are given in brackets.
• There are 100 marks available for this paper.
• You are allowed to use a calculator.
• You should use good English and present your answers in a clear and organised way.

Advice to candidates
For multiple-choice questions:
• Clearly shade the oval next to your chosen answer. For example: ⬬
• If you wish to change your answer, put a cross through your original answer.
  For example: ⊗
• If you wish to change your answer to one that you have previously crossed out,
  draw a circle around the answer. For example: ⊗

Turn over ▶

Section A: Core Technical Principles
Answer **all** the questions in this section.

1   Which **one** of the following is a type of ferrous metal?

    A   Low carbon steel

    B   Aluminium

    C   Copper

    D   Brass

[1 mark]

2   Which **one** of the following is an approach to manufacturing designed to minimise the amount of resources used and waste produced?

    A   Computer aided manufacturing

    B   Lean manufacturing

    C   Computer aided design

    D   Flexible manufacturing

[1 mark]

3   Which **one** of the following describes a material's ability to be drawn into a wire?

    A   Malleability

    B   Fusibility

    C   Ductility

    D   Electrical conductivity

[1 mark]

4   Products can be redesigned in response to market pull.
Which **one** of the following describes market pull?

    A   New technology making a product cheaper to manufacture

    B   Making a product based on the wants and needs of consumers

    C   A new material allowing a product to be lighter

    D   Making a product more expensive to buy

[1 mark]

5 **Figure 1** shows a pair of jeans, which are made from a denim.

Figure 1

What type of fibre is used to make denim?

A Cotton ⬭

B Wool ⬭

C Elastane ⬭

D Polyester ⬭

[1 mark]

6 **Figure 2** shows a gear train. What is the gear ratio of this mechanism?

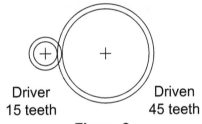

Driver
15 teeth

Driven
45 teeth

Figure 2

A 1 : 10 ⬭

B 1 : 30 ⬭

C 3 : 1 ⬭

D 60 : 1 ⬭

[1 mark]

7 What is the definition of a microcontroller?

A A small remote-controlled system ⬭

B A timer with memory ⬭

C A programmable integrated circuit (IC) with a processor and memory ⬭

D A counter made up of a series of logic gates ⬭

[1 mark]

Turn over ▶

8   Which **one** of the following statements is **true**?

A   Ink jet card is designed to let the ink bleed when used with an ink jet printer.   ◯

B   Solid white board is bleached white to make it suitable for printing on.   ◯

C   Isometric grid paper has grid squares printed onto it to make it suitable for orthographic and scale drawings.   ◯

D   Cartridge paper has a textured surface that can only be drawn on in pencil.   ◯

[1 mark]

9   **Figure 3** shows a cardboard net for sandwich packaging.

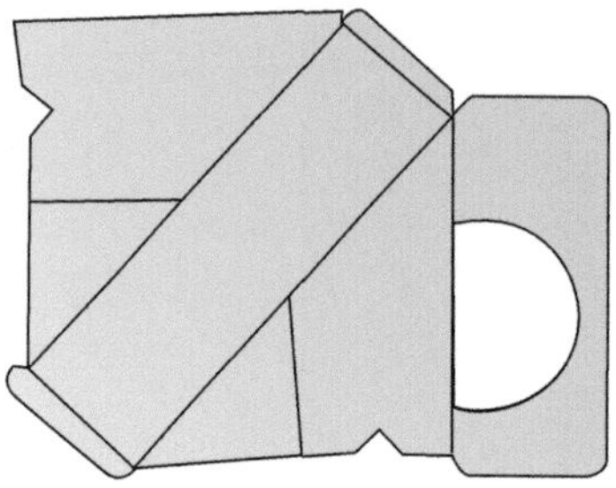

Figure 3

Which **one** of the following CAM machines could be involved in the manufacture of this product?

A   CNC router   ◯

B   Laser cutter   ◯

C   3D printer   ◯

D   CNC milling machine   ◯

[1 mark]

10   Which **one** of the following changes to a product is most likely to be damaging to the environment?

A   Making a product more durable.   ◯

B   Changing the design of a product so it can be taken apart for repair.   ◯

C   Changing a component in the product to a newer version which gives higher performance but is less reliable.   ◯

D   Making the product from materials that are all recyclable.   ◯

[1 mark]

11   State **two** properties of oak that make it suitable for use in flooring.

1.   ....

     ....

2.   ....

     ....

[2 marks]

12   Carbonfibre reinforced plastic (CRP) is an example of a composite material.
It can be used to make bulletproof vests.

Give **two** properties of CRP that make it a suitable material for this purpose.

1.   ....

     ....

2.   ....

     ....

[2 marks]

13   A new design of toothbrush is going to be made from a bioplastic.
This plastic is made from plant-based materials.

Explain why using a bioplastic to make toothbrushes is a more environmentally friendly choice
of material than using oil-based plastic.

     ....

     ....

     ....

     ....

[2 marks]

Turn over ▶

14 The bar chart in **Figure 4** shows the electricity generated from renewable and non-renewable energy sources in a small country over 20 years.

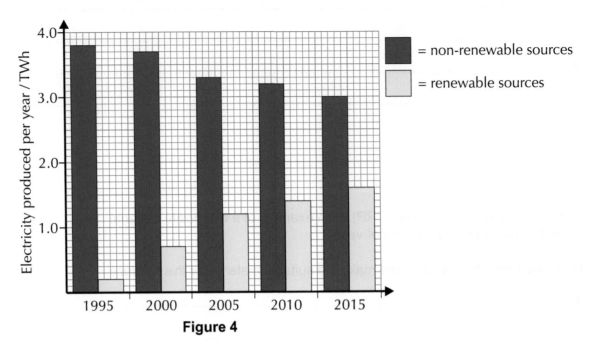

Figure 4

14.1 State the trend in the amount of electricity generated from renewable sources in **Figure 4**.

....

....
[1 mark]

14.2 Suggest **one** reason for the trend you identified in question **14.1**.

....

....

....
[1 mark]

14.3 Calculate how much **more** electricity the country produced per year in 2015 than in 1995.

....

....

....
[2 marks]

[Total 4 marks]

Section B: Specialist Technical Principles
Answer **all** the questions in this section.

15 **Figure 5** shows a range of different items.

Plastic packaging tray

Cotton fabric

Metal cogs in a watch mechanism

Printed wallpaper

Figure 5

15.1 Choose **one** of the items shown in **Figure 5**. ....

Name a process that is used to manufacture your chosen item.

Process: ....

In the box below, use sketches and/or notes to give a detailed description of this manufacturing process.

[5 marks]

Question 15 continues on the next page

Turn over ▶

15.2 For the item you have chosen in question **15.1**, describe in detail the main processes involved in the **extraction** of the raw material that it's made from, and its conversion to a useful form.

....

....

....

....

....

....

....

....

....

....

....

....

....

....

....

[4 marks]

[Total 9 marks]

16 State whether one-off, batch, mass or continuous production would be the best method of production in the following situations. Explain your answer in each case.

16.1 Production of 150 double bed frames and 200 single bed frames.

....

....

[2 marks]

16.2 Production of a 5-door family car.

....

....

[2 marks]

[Total 4 marks]

17 Standard components are frequently used in the manufacture of products.

Choose **one** of the products in the box below.

Paper catalogue	Wooden cupboard	Coat	Bicycle

Name a suitable standard component that could be used in your chosen product.

Product: ....

Standard component: ....

In the box below, use sketches and/or notes to give a detailed description of how the standard component is used.

[4 marks]

Turn over ▶

18 For **one** of the items listed below, name a suitable finish or treatment that could be applied. Give a brief description of how your chosen finish/treatment is applied and a reason for your choice of finish/treatment.

> - A book cover
> - A garden shed
> - The metal handle of a tool
> - A 2000-metre roll of plain cotton
> - A printed circuit board (PCB) in an air conditioning unit located outdoors

Chosen item: ....

Finish/treatment: ....

Description: ....

....

....

....

....

Reason for choice: ....

....

....

....

[3 marks]

19   Some products are designed so they can be reused many times.
Other products are designed for a small number of uses or even a single use.

Evaluate why products designed to be reused are considered more environmentally friendly than products designed to be used once.

....

....

....

....

....

....

....

....

....

....

....

....

....

....

....

....

....

....

....

....

....

....

....

....

....

....

[10 marks]

Turn over ▶

Section C: Designing and Making Principles
Answer **all** the questions in this section.

The product shown in **Figure 6** is a compact digital camera.
It is designed for keen adult photographers to use in a wide range of situations.

Figure 6

Specification:
- Camera features include: 20 × zoom, full HD video recording, GPS (for recording location information)
- Large 3-inch, high-resolution touchscreen
- Water and shock resistant
- Strap included (not shown)

20 Evaluate the camera in terms of its:

20.1 aesthetics

....

....

....

....

....

....

....

....

[4 marks]

20.2 ergonomics

....

....

....

....

....

....

....

[4 marks]

20.3 suitability for users

....

....

....

....

....

....

....

[4 marks]

[Total 12 marks]

Turn over for the next question

21 Study the photo of the compact camera in **Figure 6** on **page 170**.

You have been asked to redesign the camera for elderly people.
It should be comfortable for them to use and easy to operate.

21.1 Suggest **four** additions or alterations that you would make to the camera's design specification.
Explain why each one would be appropriate for an elderly target market.

1. ....
....
....
....
....

2. ....
....
....
....
....

3. ....
....
....
....
....

4. ....
....
....
....
....

[8 marks]

21.2 State an existing feature listed in the specification of the camera on **page 170** that is well suited to an elderly user. Give an explanation for your choice.

....

....

....

....
[2 marks]

21.3 You have been asked to produce a manufacturing specification for the camera.

Explain what a manufacturing specification is.
Include **two** examples of details that this specification could contain.

Explanation: ....

....

....

Example 1: ....

....

Example 2: ....

....
[3 marks]
[Total 13 marks]

Turn over for the next question

22 Identify **two** environmental considerations when designing a camera.
Explain how each one has an impact on the environment.

1. ....
....
....
....

2. ....
....
....
....

[4 marks]

23 Market research was carried out on the new camera design mentioned in
Question 21, to find out the likes and dislikes of some potential users.
The research was carried out using questionnaires given to a group of 360 elderly people.
The results of the market research are shown in **Figure 7**.

Q1 What amount of money would you expect this camera to be worth?

Price range	Number of people
£100-125	27
£126-150	41
£151-175	83
£176-200	132
£201-225	47
£226-250	25
More than £251	5
Total	360

Note: the specification of the camera was shown on the questionnaire.

Q2 What is the minimum number of photos you would want the camera to store?

Number of photos	Number of people
Less than 400	38
401-800	69
801-1200	110
1201-1600	98
More than 1601	45
Total	360

Q3 What type of material would you prefer the main body of the camera to be made from?
Metal: 20%   **Plastic: 55%**
Don't mind: 25%

Figure 7

23.1 Using the results of **Q3** shown in **Figure 7**, calculate the number of people who answered 'metal' and the number of people who answered 'plastic'.

....

....

....

....

....

....
[1 mark]

23.2 Using the results of **Q1** in **Figure 7**, calculate the percentage of people who said they would expect to pay £200 or less for the camera. Give your answer to **1 decimal place**.

....

....

....

....
[1 mark]

Question 23 continues on the next page

Turn over ▶

23.3 Describe how the results of the market research shown in **Figure 7** should affect the design of the camera. Discuss the results of each question and use data to support your answers.

Q1 results: ....

....

....

....

....

....

Q2 results: ....

....

....

....

....

....

Q3 results: ....

....

....

....

....

....

[6 marks]

[Total 8 marks]

24 **Figure 8** shows the net of a gift box.

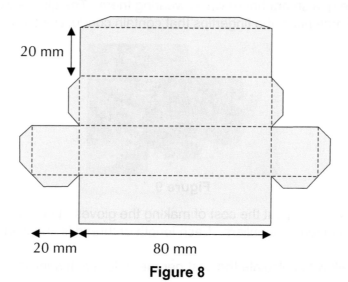

20 mm

20 mm   80 mm

Figure 8

24.1 On the isometric grid below, produce an isometric drawing of the gift box shown in **Figure 8** when it is assembled.

10 mm

[2 marks]

24.2 Each dimension of the gift box has a tolerance of ± 1.5 mm.
The assembled gift box has a height of 21.2 mm, a width of 18.4 mm and a length of 79.1 mm.
Explain why this box does **not** fall within the stated tolerance.

....

....

[2 marks]
[Total 4 marks]

Turn over ▶

25 **Figure 9** shows a pair of gloves for use in cold conditions that allow the wearer to use a touchscreen device (e.g. a smartphone) whilst wearing them. The gloves work because they have patches of material in the fingertips that contain conductive thread.

Figure 9

A clothing company is working out the cost of making the gloves. Lengths of conductive thread are available from several different suppliers. Each length of thread is supplied on a plastic cone.

25.1 Complete the table below to calculate the cost per metre for each supplier.

[1 mark]

Supplier	Length of thread per cone (m)	Cost of cone (£)	Cost per metre (£)
A	1650	246.50	0.15
B	1000	40.00	
C	2250	202.50	

25.2 One of the designs for the gloves uses 65 cm of conductive thread in each pair.
Calculate the total cost of thread that would need to be purchased to make 7000 pairs of gloves.
The thread will be purchased from supplier A and can be bought in whole cones only.

....

....

....
[3 marks]
[Total 4 marks]

26.1 Give the design strategy that is centred around a constant process of evaluation and improvement.

....
[1 mark]

26.2 Collaborating with other people is an important part of the design process.
State **two** groups of people that a designer could collaborate with.
Explain how collaboration could be useful in each case.

1. ....

....

....

2. ....

....

....
[4 marks]
[Total 5 marks]

END OF QUESTIONS

Section One — Key Ideas in Design and Technology

Page 14 (Warm-Up Questions)

1   E.g. it reduces the amount of space needed for the storage of materials/finished products. It means there's less money tied up in materials that aren't being used.
2   The bigger a product's carbon footprint is, the larger its contribution to global warming.
3   When products are designed to become obsolete/ useless quickly.
4   Repair, re-use, recycle, rethink, reduce, refuse
5   continuous improvement
6   A non-renewable energy resource will one day run out but a renewable energy resource can be renewed as it is used.

Pages 15-16 (Exam Questions)

1   C *[1 mark]*
2   C *[1 mark]*
3   B *[1 mark]*
4 a)   Any two from: e.g. it can encourage consumers to replace their existing power tools with new ones, which can lead to older models being disposed of, which can cause pollution. / New, replacement power tools that are made have a carbon footprint. / New power tools being manufactured, packaged, transported and eventually disposed of can result in the increased usage of finite resources/environmental damage linked to the collection of resources *[2 marks]*.
   b)   E.g. newer power tools are likely to have more efficient components than older power tools *[1 mark]*. This means they may have lower carbon footprints *[1 mark]*.
5 a)   Any two from: e.g. robots can increase the speed of production as they can work faster than humans and don't need to rest. / Robots work faster than humans so they can be cheaper to use than human workers. / Robots can increase the quality of manufacture as they work with high accuracy, whereas humans can make mistakes. / Robots can reduce costs as they work with high accuracy, whereas humans can make mistakes (which have costs, e.g. time and materials). / Robots can be used in dangerous situations where it would be unsafe for humans *[2 marks]*.
   b)   E.g. they can't carry out tasks that require human judgement / can be very expensive to buy *[1 mark]*.
6 a)   E.g. it means they can be reused rather than being disposed of, which might cause pollution *[1 mark]*.
   b)   E.g. the envelopes will rot away naturally when disposed of but the bubble wrap will not *[1 mark]*. This means that the bubble wrap will permanently take up space in landfill *[1 mark]*.
   c)   E.g. the padded envelope can't be recycled as a whole *[1 mark]*. The bubble wrap and paper need to be separated first, which can be difficult *[1 mark]*.

Section Two — An Introduction to Materials and Systems

Page 28 (Warm-Up Questions)

1   Plastics are good electrical insulators.
2   The ink spreads out.
3   E.g. it's ductile and a good electrical conductor.
4   Any two from: natural fibres are biodegradable / recyclable / made from renewable materials.
5   E.g. warp-knitted fabrics keep their shape but weft-knitted fabrics can lose their shape / warp-knitted fabrics are less likely to 'ladder' than weft-knitted fabrics.

Page 29 (Exam Questions)

1   B *[1 mark]*
2   Property: e.g. high strength-to-weight ratio / low density / soft *[1 mark]*
      Use: e.g. modelling *[1 mark]*

3   A blended fabric is made from a yarn that is a combination of two or more different types of fibre *[1 mark]*. A mixed fabric is made from two or more different types of yarn *[1 mark]*.
4 a)   Plastic — e.g. polyvinyl chloride / PVC *[1 mark]*.
       Reason — e.g. it's cheap so it can be used in mass production / it's durable so offers long-lasting protective packaging *[1 mark]*.
   b)   Plastic — e.g. polyethylene terephthalate / PET *[1 mark]*
       Reason — e.g. it's light so is good for being carried around / it's strong so will withstand being knocked or squashed without breaking / it's tough so will bend a little rather than breaking *[1 mark]*.

Page 40 (Warm-Up Questions)

1   Boards with thin copper tracks that connect components in a circuit.
2   The output will be on if both of the two inputs are on, otherwise the output will be off.
3   For half a turn the follower won't move, then it will gently rise and fall.
4   silver

Page 41 (Exam Questions)

1   A *[1 mark]*
2   C *[1 mark]*
3   velocity ratio = 105 ÷ 35 = 3/1 / 3:1 / 3 *[1 mark]*
4 a)   E.g. shape memory alloy/nitinol / photochromic pigments *[1 mark]*.
   b)   E.g.
       Shape memory alloy/nitinol:
       If you deform products made from this, they can be returned to their original shape by heating *[1 mark]*, so frames made from this can be easily fixed if they get accidentally bent out of shape *[1 mark]*.
       Photochromic pigment:
       It can change colour when exposed to different levels of light *[1 mark]*, so sunglasses with photochromic lenses can be designed to get darker in bright light, and clearer in low light *[1 mark]*.

Section Three — More about Materials

Page 51 (Warm-Up Questions)

1   A duty to act in a way that benefits society and the environment.
2   torsion
3   Down time is the time between batches when you're not making anything. During this time, machines and tools may have to be set up differently/changed for the next batch.
4 a)   mass production
   b)   It's easy to recruit people because they don't have to be highly skilled.
5 a)   To check products have been made to a high enough standard and to make sure they meet the manufacturing specification.
   b)   E.g. it would take too long to test every product/component that was manufactured.

Page 52 (Exam Questions)

1 a)   Any two from: e.g. the material needs to be strong enough to support the weight of items that will be put on it / it needs to be lightweight so the table is portable / it must be able to withstand outdoor conditions *[2 marks]*.
   b)   E.g. materials that are widely available are usually less expensive / quicker and easier to source *[1 mark]*.
   c)   E.g. buying the materials in bulk allows the company to negotiate a discount with the supplier *[1 mark]*. This means the table can be made for less money, so can be sold for a cheaper price *[1 mark]*.

2 a) E.g. a depth stop is a long rod that is clamped close to the drill bit *[1 mark]*. Once the chosen depth has been reached, the depth stop comes into contact with the material and prevents the drill from going any deeper *[1 mark]*.

b) E.g. the power settings *[1 mark]* and the feed rate *[1 mark]*.

c) E.g. keeping the PCB exposure times to UV light/the developer solution/the etching solution constant *[1 mark]*.

Page 59 (Warm-Up Questions)

1 the point where x, y and z meet / the datum (0,0,0)

2 A template is drawn round with a pencil/cut round with a knife.

3 ores

Page 60 (Exam Question)

1 a) Cotton — cotton plant *[1 mark]*
Polyester — crude oil *[1 mark]*

b) i) E.g. farming for cotton fibres often uses artificial fertilisers/pesticides, which can pollute rivers and harm wildlife *[1 mark]*. Land may need to be cleared for farming, which could involve deforestation/clearing of habitats *[1 mark]*.

ii) Any two from: e.g. drilling for crude oil can result in toxic chemicals being released into the atmosphere / can result in waste material/oil leaks, which pollute the surrounding habitats / may require land to be cleared to make room for the drill site, which can destroy habitats *[2 marks]*.

c) How to grade your answer:
[No marks] There is no relevant information.
[1 mark] There is a brief description of the process, but key stages are left out and the answer contains a number of errors. AND/OR there is a diagram but it lacks detail and clarity.
[2 marks] There is a description of the process, but some points are missing or there are some errors. AND/OR there is a diagram with some annotations, but it lacks detail or contains errors.
[3 marks] There is a detailed description of the process, with most stages in the correct order but the description may contain small errors or lack some clarity. AND/OR there is an annotated diagram, which is mainly correct but some points are missing.
[4 marks] There is a clear, accurate and detailed description of the process, including the key stages in the correct order. AND/OR there is an accurate and appropriately annotated diagram clearly showing the process.
Here are some points your answer may include:
Cotton:
The plants are treated with chemicals to make the leaves fall off.
The cotton fibres are then harvested.
The cotton fibres are cleaned to remove dirt.
The seeds are removed from the pods.
The fibres are then combed using wire rollers (carding).
The fibres are then spun into yarn.
Polyester:
Crude oil is fractionally distilled.
Certain fractions are polymerised to make a polymer (polyester).
The polymer is then melted and forced through tiny holes to form long filaments.
The filaments are left to cool, before being spun into yarn.
(Relevant, labelled sketches with annotations showing these points should also be credited.)

Section Four — Paper and Board

Page 72 (Warm-Up Questions)

1 true

2 Any two from: e.g. staples/tabs/treasury tags/Velcro® pads/drawing pins/prong paper fasteners

3 e.g. a die cutter

4 laminating

Page 73 (Exam Questions)

1 Using standard components saves time during manufacture / is more efficient *[1 mark]*. Using standard components means that specialist machinery and extra materials aren't needed, which saves money *[1 mark]*.

2 a) Quote 1 = 180 ÷ 100 = £1.80
Quote 2 = 340 ÷ 100 = £3.40 *[1 mark for two correct prices]*

b) £3.40 − £1.80 = £1.60 *[1 mark]*
The cost of embossing each invitation can be calculated by subtracting the cost per invitation of quote 2 (which includes embossing) from the price per invitation of quote 1 (which includes all features except embossing).

c) $((340 - 180) \div 180) \times 100 = 88.9\%$ *[2 marks for an answer of 88.9 — 1 mark for correct working but incorrect final answer]*
To work out the percentage increase you need to divide the difference between the two quotes (the amount that quote 2 is higher than quote 1) by the original number (quote 1) and multiply the result by 100.

3 How to grade your answer:
[No marks] There is no relevant information.
[1-2 marks] Brief consideration of the positive and negative aspects of each property with regards to packaging. Limited or no conclusion drawn. Points discussed may be limited to only positive OR negative aspects.
[3-4 marks] Positive AND negative aspects of each property discussed and supported with sensible explanation. Conclusion drawn after considering positive and negatives aspects of each property.
Here are some points your answer may include:
Rigidity:
Packaging needs to be rigid so that it keeps its shape and gives effective protection to what is inside.
Rigid packaging is more difficult to form, as it is hard to bend/fold it into a 3D shape (e.g. a cardboard box).
Strength:
Packaging needs to be strong so it can withstand a fair amount of force before bending and/or breaking.
Strong packaging is usually heavier/bulkier, which can increase transport costs.

Section Five — Wood, Metals and Polymers

Page 83 (Warm-Up Questions)

1 e.g. urea formaldehyde

2 e.g. granules / powders

3 A temporary fitting that allows furniture to be assembled and taken apart easily.

4 e.g. a coping saw

Page 84 (Exam Questions)

1 a) e.g. rip saw / tenon saw *[1 mark]*

b) e.g. so an even pressure can be applied / to give better grip *[1 mark]*

c) E.g. the bradawl creates a dent *[1 mark]*, which helps you to drill in the right place / stop the drill from slipping *[1 mark]*.

d) e.g. bench plane *[1 mark]*

2 How to grade your answer:
[No marks] There is no relevant information.
[1 mark] There is a brief description of the method, but key stages are left out and the answer contains a number of errors. AND/OR there is a diagram but it lacks detail and clarity.
[2 marks] There is a description of the method, but some points are missing or there are some errors. AND/OR there is a diagram with some annotations, but it lacks detail or contains errors.
[3 marks] There is a detailed description of the method, with most stages in the correct order but the description may contain small errors or lack some clarity. AND/OR there is an annotated diagram, which is mainly correct but some points are missing.

[4 marks] There is a clear, accurate and detailed description of the method, including the key stages in the correct order. AND/OR there is an accurate and appropriately annotated diagram clearly showing the method.
Here are some points your answer may include:
A rivet is a metal peg with a head at one end.
A hole is drilled through both pieces of metal.
The rivet is inserted into the hole with a set.
The head is held against the metal whilst the other end is flattened and shaped into another head using a hammer.
(Relevant, labelled sketches with annotations showing these points should also be credited.)

Page 93 (Warm-Up Questions)

1   E.g. it can be used to make straight or curved cuts in all materials.
2   e.g. a sheet metal folder
3   A sheet of thermoforming plastic is heated until it softens. The sheet is placed on a mould and left to cool. The mould is removed and the plastic stays in the shape of the mould.
4   galvanisation

Page 94 (Exam Questions)

1   How to grade your answer:
     [No marks] There is no relevant information.
     [1 mark] There is a brief description of the process, but key stages are left out and the answer contains a number of errors. AND/OR there is a diagram but it lacks detail and clarity.
     [2 marks] There is a description of the process, but some points are missing or there are some errors. AND/OR there is a diagram with some annotations, but it lacks detail or contains errors.
     [3 marks] There is a detailed description of the process, with most stages in the correct order but the description may contain small errors or lack some clarity. AND/OR there is an annotated diagram, which is mainly correct but some points are missing.
     [4 marks] There is a clear, accurate and detailed description of the process, including the key stages in the correct order. AND/OR there is an accurate and appropriately annotated diagram clearly showing the process.
     Here are some points your answer may include:
     The metal is melted.
     It is then poured into a mould/die, which is in the shape of the model car's body.
     The metal is left to cool and solidify.
     The body of the model car can be removed from the mould and trimmed to remove any excess material.
     (Relevant, labelled sketches with annotations showing these points should also be credited.)
2   Moulding process: Plastic bottle — blow moulding / Plastic guttering — extrusion *[1 mark]*
Plastic bottles are hollow, so blow moulding is a suitable choice. Plastic guttering is a long continuous strip with the same cross-section throughout, so extrusion is a suitable moulding process to use.
     How to grade your answer:
     [No marks] There is no relevant information.
     [1 mark] There is a brief description of the process, but key stages are left out and the answer contains a number of errors. AND/OR there is a diagram but it lacks detail and clarity.
     [2 marks] There is a description of the process, but some points are missing or there are some errors. AND/OR there is a diagram with some annotations, but it lacks detail or contains errors.
     [3 marks] There is a detailed description of the process, with most stages in the correct order but the description may contain small errors or lack some clarity. AND/OR there is an annotated diagram, which is mainly correct but some points are missing.
     [4 marks] There is a clear, accurate and detailed description of the process, including the key stages in the correct order. AND/OR there is an accurate and appropriately annotated diagram clearly showing the process.

Here are some points your answer may include:
Blow moulding:
This process starts with a tube of softened plastic.
The plastic is inserted into a solid mould.
Air is injected into the tube of plastic, forcing the plastic to expand.
The plastic takes the shape of the inside of the mould.
The mould is opened to remove the plastic bottle.
Extrusion:
The plastic is heated in a chamber in the machine until it melts.
The liquid plastic is then forced through a mould/die under pressure.
This produces a long continuous strip of plastic/guttering.
The guttering will have the same cross-section as the shape of the exit hole.
(Relevant, labelled sketches with annotations showing these points should also be credited.)

Section Six — Textiles

Page 108 (Warm-Up Questions)

1   lamination
2   One half of the Velcro® has rough nylon hooks which attach to the soft loops on the other half.
3   To join the edges of fabric pieces together securely.
4   E.g. large amounts of fabric can be dyed at once / fabric can be dyed quickly.
5   e.g. a squeegee and a screen

Pages 109-110 (Exam Questions)

1 a)   Fabric A: total cost = 5.50 × 1.70 = £9.35
     Fabric B: total cost = 7.40 × 1.55 = £11.47
     [1 mark for two correct total costs]
   b) i)   Total cost = 6.20 × 1.90 = £11.78 *[1 mark]*
     ii)   Area of fabric C required for the original design
     = 1.50 × 1.55 = 2.325 m² *[1 mark]*
     Area of fabric C required for the revised design
     = 1.50 × 1.90 = 2.85 m² *[1 mark]*
The width of the fabric is given in cm so this needs to be converted to metres in these calculations, i.e. 155 cm ÷ 100 = 1.55 m.
     Difference in area = 2.85 – 2.325 = 0.525 m²
     = 0.53 m² (to 2 decimal places) *[1 mark]*
2 a) i)   E.g. when doing a lot of sewing *[1 mark]* because using a machine is much faster than hand-sewing *[1 mark]*.
     ii)   E.g. for small tasks like embroidery or darning *[1 mark]* because hand-sewing is more precise *[1 mark]*.
   b)   E.g. to check that the thread tension / stitch length / stitch type is correct *[1 mark]*.
3   Any two from: e.g. they're warm/soft to walk on / they're resistant to fading *[2 marks]*
4 a)   How to grade your answer:
     [No marks] There is no relevant information.
     [1 mark] There is a brief description of the method, but key stages are left out and the answer contains a number of errors. AND/OR there is a diagram but it lacks detail and clarity.
     [2 marks] There is a description of the method, but some points are missing or there are some errors. AND/OR there is a diagram with some annotations, but it lacks detail or contains errors.
     [3 marks] There is a detailed description of the method, with most stages in the correct order but the description may contain small errors or lack some clarity. AND/OR there is an annotated diagram, which is mainly correct but some points are missing.
     [4 marks] There is a clear, accurate and detailed description of the method, including the key stages in the correct order. AND/OR there is an accurate and appropriately annotated diagram clearly showing the method.

Here are some points your answer may include:
The printing block has a raised design.
The design is star-shaped.
Printing ink is applied to the raised surface of the block.
The block is pressed down onto the fabric to transfer the star design to the fabric.
(Relevant, labelled sketches with annotations showing these points should also be credited.)

b) Any two from, e.g. different colours (for the different stars) can be printed using the same printing block / the design can be easily repeated / the block won't wear out easily *[2 marks]*.

Section Seven — Electronic and Mechanical Systems

Page 120 (Warm-Up Questions)

1 If the car is involved in an accident, these properties will help the car body absorb the impact of the crash, protecting the passengers.
2 The current and voltage values that an electronic component is designed to work at.
3 The casing holds/protects the IC. The pins connect the IC to the rest of the circuit/PCB.
4 It acts as an electrical connection between the component's pins and the copper tracks on the PCB.
5 It is the process of placing components onto a PCB in the correct position and orientation.
6 a) polymers/plastic
  b) E.g. it acts as a protective barrier / it helps to decrease corrosion/increase the durability of the board.

Page 121 (Exam Questions)

1 a) 10% of 56 ohms = 5.6 ohms *[1 mark]*
   lower limit of the range = 56 − 5.6 = 50.4 ohms *[1 mark]*
   upper limit of the range = 56 + 5.6 = 61.6 ohms *[1 mark]*
  b) blue, grey, orange *[1 mark]*
The first two bands give the first two digits of the resistance value so you just need to look up their numbers in the table to see what colours they represent (i.e. blue, grey). Band 3 gives the number of zeros that comes after this (3 in this example), so this band is orange.

2 Name: flow/wave soldering *[1 mark]*
How to grade your answer:
[No marks] There is no relevant information.
[1 mark] There is a brief description of the method, but key stages are left out and the answer contains a number of errors. AND/OR there is a diagram but it lacks detail and clarity.
[2 marks] There is a description of the method, but some points are missing or there are some errors. AND/OR there is a diagram with some annotations, but it lacks detail or contains errors.
[3 marks] There is a detailed description of the method, with most stages in the correct order but the description may contain small errors or lack some clarity. AND/OR there is an annotated diagram, which is mainly correct but some points are missing.
[4 marks] There is a clear, accurate and detailed description of the method, including the key stages in the correct order. AND/OR there is an accurate and appropriately annotated diagram clearly showing the method.
Here are some points your answer may include:
Components are placed on the board.
Components are sometimes glued in place.
The board is passed over a pan of molten solder.
A 'wave'/upwelling of solder is produced by a pump.
The wave produced is at the right height to just touch the base of the board as it is passed over the pan, soldering the components in place.
(Relevant, labelled sketches with annotations showing these points should also be credited.)

Section Eight — Designing and Making

Page 133 (Warm-Up Questions)

1 E.g. chest size / collar size / height.
2 client
3 A list of conditions that the proposed product should meet.

Page 134 (Exam Questions)

1 a) E.g. the coat has a water-resistant finish, which makes it suitable for wet weather *[1 mark]*. However, it isn't padded so it might not be warm enough for winter *[1 mark]*.
  b) E.g. the coat is made out of nylon/the buttons are made out of plastic, which is cheap to buy *[1 mark]*. However, the metal brooch will increase the cost of the coat *[1 mark]*.
  c) E.g. nylon/plastic is made from fossil fuels, which are finite resources/will eventually run out, so it isn't a sustainable material *[1 mark]*. However, the plastic buttons have been recycled, which makes it more environmentally friendly *[1 mark]*.
  d) E.g. bad/unsafe working conditions / paying workers unfairly *[1 mark]*.

2 How to grade your answer:
[No marks] There is no relevant information.
[1 to 2 marks] There is a brief comparison of the work of two named designers. Discussion of differences and/or similarities between the work of the two designers is vague and the answer contains major errors. Key points about the work of the two designers are missing.
[3 to 4 marks] There is some comparison of the work of the two named designers. Similarities and/or differences given are limited and errors are present.
[5 to 6 marks] There is a good attempt to compare the work of the two named designers. Similarities and differences are discussed. The answer may feature examples to help describe the work of each designer. The answer may lack clarity or contain some minor errors.
[7 to 8 marks] There is a clear, detailed and accurate comparison of the work of the two named designers. Similarities and differences between the work of the designers are discussed thoroughly using examples. There are few, if any, errors in the answer.
This is a big mark question. To get high marks you need to talk about the styles of the designers you've chosen and then compare these. Using examples is a good way to break into the higher marks — this lets you talk about specifics rather than relying on vaguer statements about the sort of work each designer is known for. There's a huge variety of things that you could say to answer this question, so we've just given advice on how marks would be awarded rather than any specific points to include.

Page 143 (Warm-Up Questions)

1 design fixation
2 This is where you check the model against the design specification.
3 E.g. advantage: it's easier to get the dimensions right with isometric drawings.
E.g. disadvantage: isometric drawings don't show things smaller the further away they are, so they look less realistic.

Page 144 (Exam Questions)

1 How to grade your answer:
[No marks] There is no relevant information.
[1 mark] There is a brief description of the process, but key stages are left out and the answer contains a number of errors. AND/OR there is a diagram but it lacks detail and clarity.
[2 marks] There is a description of the process, but some points are missing or there are some errors. AND/OR there is a diagram with some annotations, but it lacks detail or contains errors.

[3 marks] There is a detailed description of the process, with most stages in the correct order but the description may contain small errors or lack some clarity. AND/OR there is an annotated diagram, which is mainly correct but some points are missing.

[4 marks] There is a clear, accurate and detailed description of the process, including the key stages in the correct order. AND/OR there is an accurate and appropriately annotated diagram clearly showing the process.

Here are some points your answer may include:

Develop a design brief and specification.

Sketch and model design ideas.

Make a prototype.

Hold a focus group for the prototype to be tested by the target market / test the prototype yourself.

Evaluate the design / identify any problems based on testing/ feedback.

Improve the design to fix any problems.

Make another prototype.

Continue making new and improved prototypes by repeating the same method until all the problems have been identified and fixed.

(Relevant, labelled sketches with annotations showing these points should also be credited.)

2 a) Name: exploded diagram *[1 mark]*

Explanation: because it shows how the parts of a product fit together *[1 mark]*.

b) E.g.

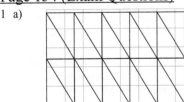

Accurate reproduction of the bookcase exploded *[1 mark]*. Each part of the bookcase is in line with the part it's attached to *[1 mark]*. Dotted lines show where parts explode from *[1 mark]*.

Page 153 (Warm-Up Questions)

1 Prototypes are full-size, fully-functioning products or systems made to test the product/system and its production methods.

2 Finding an efficient arrangement of shapes to minimise waste during cutting and shaping.

3 Standard components have already been tested for safety by the manufacturer, so you know they're likely to be safe to use in the product.

Page 154 (Exam Questions)

1 a)

[1 mark]

b) Area = ½ × 15 × 24 = 180 cm² *[1 mark]*

Area of a triangle = ½ × width × height.

c) Area of sheet = 80 × 50 = 4000 cm² *[1 mark]*

Area of 20 triangles = 180 × 20 = 3600 cm² *[1 mark]*

So material wasted = 4000 − 3600 = 400 cm² *[1 mark]*

2 The label does show the name of the product *[1 mark]* but not what it smells of *[1 mark]*. It has an old-fashioned rather than futuristic/scientific appearance *[1 mark]*. It doesn't include an image of glossy hair *[1 mark]*.

Practice Paper

Section A: Core Technical Principles
Pages 160-164

1 A *[1 mark]*
2 B *[1 mark]*
3 C *[1 mark]*
4 B *[1 mark]*
5 A *[1 mark]*
6 C *[1 mark]*
7 C *[1 mark]*
8 B *[1 mark]*
9 B *[1 mark]*
10 C *[1 mark]*
11 Any two from: e.g. it has attractive grain markings / it finishes well / it is durable / it is tough / it is very strong *[2 marks]*.
12 Any two from: e.g. it is light / it is tough / it is strong *[2 marks]*.
13 E.g. when plant materials are used to make the bioplastic more can be planted to replace them/plant material is a non-finite resource *[1 mark]* whereas, oil-based plastic is made from a finite resource *[1 mark]*. / When the bioplastic toothbrush is disposed of at the end of its life it can breakdown fully *[1 mark]* rather than taking up space in landfill like an oil-based plastic toothbrush *[1 mark]*.

Remember that finite resources can also be called non-renewable resources and non-finite resources can be called renewable resources. It's easy to confuse them, so double check you've written the one you meant to if you mention them in an answer to an exam question.

14.1 E.g. from 1995 to 2015, the production of electricity from renewable resources increased, from 0.2 to 1.6 TWh *[1 mark]*.

14.2 E.g. extracting and burning fossil fuels damages the environment. / Many people think it's better to learn to get by without non-renewables before they run out. / Improved efficiency in renewable power production mean renewables are becoming a more attractive option. / Governments have begun to introduce targets for using more renewable resources, and for cutting down on carbon dioxide emissions *[1 mark]*.

14.3 2015: 3.0 + 1.6 = 4.6 TWh
1995: 3.8 + 0.2 = 4.0 TWh
4.6 − 4.0 = 0.6 TWh
[2 marks for a correct answer, otherwise 1 mark for correctly calculating the electricity produced each year.]

To get the right data from the graph to use in these calculations, you need to make sure you've worked out the scale of the y-axis (the vertical axis). In this question each square on the y-axis is equal to 0.1 TWh because there are 10 squares for every 1.0 TWh (1.0 ÷ 10 = 0.1). For example, this means that six squares up from 1.0 is equal to 1.6 TWh.

Section B: Specialist Technical Principles
Pages 165-169

15.1 Process:

Plastic packaging tray — e.g. vacuum forming

Cotton fabric — e.g. plain weaving

Metal cogs in a watch mechanism — e.g. laser cutting

Printed wallpaper — e.g. flexography *[1 mark]*

How to grade your answer:

[No marks] There is no relevant information.

[1 mark] There is a brief description of the process, but key stages are left out and the answer contains a number of errors. AND/OR there is a diagram but it lacks detail and clarity.

[2 marks] There is a description of the process, but some points are missing or there are some errors. AND/OR there is a diagram with some annotations, but it lacks detail or contains errors.

[3 marks] There is a detailed description of the process, with most stages in the correct order but the description may contain small errors or lack some clarity. AND/OR there is an annotated diagram, which is mainly correct but some points are missing.

[4 marks] There is a clear, accurate and detailed description of the process, including the key stages in the correct order. AND/OR there is an accurate and appropriately annotated diagram clearly showing the process.

Here are some points your answer may include:

Plastic packaging tray:

A mould of the tray is put onto a vacuum bed.

A film/sheet of thermoforming plastic is clamped above the vacuum bed.

The plastic sheet/film is heated until it goes soft.

The vacuum bed is lifted close to the heated plastic.

Air is sucked out from under the plastic, creating a vacuum and forcing the plastic onto the mould of the tray.

The moulded plastic is cooled and the vacuum bed lowered.

The cold plastic is rigid so holds the shape of the tray.

Cotton fabric:

A loom is used to weave cotton yarns into a fabric.

Cotton fabrics are woven by interlacing a weft yarn and a warp yarn.

The weft travels from right to left and the warp yarn travels up and down the weave.

The yarns are interlaced by passing the weft yarn over and under alternate warp yarns to create a plain weave.

Looms can be operated by hand or computer-controlled.

Metal cogs in a watch mechanism:

The cog is designed on a computer using computer-aided design (CAD) software.

The correct feed rate values and power settings are programmed into the laser cutter.

The type and thickness of the material being used determine the feed rate and power settings needed.

A laser beam cuts through the material to accurately cut out the exact shape of a cog.

The laser is guided using computer numeric control (CNC) — it follows the coordinates set out in the design.

Printed wallpaper:

The printing plate is made from flexible rubber/plastic.

The striped pattern sticks out a bit from the plate.

Paint is applied to the raised areas of the printing plate.

The printing plate is rolled over the wallpaper to transfer the striped pattern to the paper.

(Relevant, labelled sketches with annotations showing these points should also be credited.)

15.2   How to grade your answer:

[No marks] There is no relevant information.

[1 mark] There is a brief description of the processes involved, but key stages are left out and the answer contains a number of errors.

[2 marks] There is a description of the processes involved, but some points are missing or there are some errors.

[3 marks] There is a detailed description of the processes involved, with most stages in the correct order but the description may contain small errors or lack some clarity.

[4 marks] There is a clear, accurate and detailed description of the processes involved, including the key stages in the correct order.

Here are some points your answer may include:

Plastic packaging tray:

Drilling is used to extract crude oil from the ground.

The crude oil is transported to a refinery.

At the refinery, it is heated using a process called fractional distillation, which separates it into different chemicals called fractions.

Polymerisation is then used to link some of the fractions together to make polymers (plastics).

Some fractions may need to be heated and broken down into smaller molecules using a process called cracking, before they can become polymerised.

The plastic can then be used, e.g. by being moulded into the desired shape.

Cotton fabric:

Cotton fibres are extracted from the seed pods of the cotton plant.

Before harvesting, the plants are treated with chemicals to make the leaves fall off.

The fibres are cleaned to remove dirt, and the seeds are removed.

The cotton fibres then undergo a process called carding where they're combed using wire rollers.

The cotton fibres are then spun into yarns.

Metal cogs in a watch mechanism:

Metal ore is mined from the ground.

The metal is then extracted from the ore. Usually this is done by either crushing and heating the ore in a furnace or using electrolysis.

The metal then needs to be refined to remove impurities.

The metal is then cast/moulded into a certain shape. In this case, it would be moulded into sheets.

Printed wallpaper:

Trees are cut down and taken to a paper mill.

The bark is stripped off and the wood is then cut into small pieces by a chipper.

These small bits of wood undergo mechanical/chemical pulping to convert the wood to individual cellulose fibres/pulp.

The pulp is washed and pressed flat between rollers to form paper.

The paper is dried and cut to size.

16.1   Batch production *[1 mark]* — e.g. because the machines can be easily altered to produce the different types of bed frame *[1 mark]*.

16.2   Mass production *[1 mark]* — e.g. because thousands of identical cars will be made / the cars can be made on a production line *[1 mark]*.

17   Standard component:

Paper catalogue — e.g. comb binding

Wooden cupboard — e.g. butt hinge

Coat — e.g. zip

Bicycle — e.g. roller chain *[1 mark]*

How to grade your answer:

[No marks] There is no relevant information.

[1 mark] There is a brief description of how the named component is used, but some details are left out and the answer contains a number of errors. AND/OR there is a diagram but it lacks detail and clarity.

[2 marks] There is a detailed description of how the named component is used, but with some errors. AND/OR there is well-annotated diagram, but there are some errors.

[3 marks] There is a clear, accurate and detailed description of how the named component is used, including key details. AND/OR there is a clear, accurate and appropriately annotated diagram showing how the named component is used.

18   A book cover:

Finish/treatment: e.g. embossing *[1 mark]*

Description: e.g. a shaped die is pushed into the back of the material to leave a slightly raised impression on its surface *[1 mark]*.

Reason for choice: e.g. it's often used to draw attention to a particular bit of a product, e.g. the title of a book *[1 mark]*.

A garden shed:

Finish/treatment: e.g. tanalising *[1 mark]*

Description: e.g. timber is placed in a tank which is flooded with preservative and then pressurised to force the preservative deep into the wood *[1 mark]*.

Reason for choice: e.g. the treatment helps to prevent insect attacks and the decay of the wood, meaning it will last longer *[1 mark]*.

The metal handle of a tool:

Finish/treatment: e.g. dip coating *[1 mark]*

Description: e.g. a metal is heated evenly in an oven before being plunged into fluidised powder. It is then returned to the oven, which causes the thin layer of plastic to fuse to the surface of the handle *[1 mark]*.

Reason for choice: e.g. it offers a soft, smooth finish for tool handles *[1 mark]*.

A 2000-metre roll of plain cotton:

Finish/treatment: e.g. industrial flat-bed screen printing *[1 mark]*

Description: e.g. fabric passes under screens, each filled with a different colour of dye, on a conveyor belt and the colours are applied one after the other *[1 mark]*.

Reason for choice: e.g. this method allows very long lengths of fabric to be printed on quickly *[1 mark]*.

A printed circuit board (PCB) in an air conditioning unit located outdoors:

Finish/treatment: e.g. PCB lacquering *[1 mark]*

Description: e.g. PCB lacquer is a thin polymer film that is sprayed or painted on to a PCB, or is applied by dipping the whole board into it *[1 mark]*.

Reason for choice: e.g. the lacquer provides a protective barrier against moisture, chemicals, large temperature changes and dust, which might be encountered by the PCB as it is situated outdoors *[1 mark]*.

19 How to grade your answer:

[No marks] There is no relevant information.

[1-2 marks] Answer includes one or two valid points, made with little or no explanation. No conclusions are drawn.

[3-4 marks] Answer includes a small number of valid points, made with some explanation. There is limited evaluation. There may be a brief conclusion drawn, but it doesn't link clearly to the points described.

[5-6 marks] Answer includes several valid points, which show good understanding and are mostly explained. Points given are sometimes unclear. Conclusions are drawn, but are mostly unsupported by the points discussed.

[7-8 marks] Answer includes a range of valid points that show good understanding and are well explained. Points given are mostly clear. Conclusions are drawn, but are sometimes unsupported by the points discussed.

[9-10 marks] Answer includes a wide range of valid points that show a thorough understanding and are explained in depth. Points given are clear. Conclusions are drawn and are fully supported by the points discussed.

Here are some points your answer may include:

All products have an environmental impact during their life cycle, e.g. when they're manufactured, transported, used and eventually disposed of.

Negative environmental impacts that occur during a product's life cycle include the emission of greenhouse gases (contributing to global warming), the destruction of natural habitats and pollution.

For example, raw materials need to be obtained before products are manufactured. This may occur through mining, drilling, and cutting down trees, which can all cause pollution and the destruction of natural habitats.

The environmental impact of obtaining raw materials and manufacturing products can be reduced by using sustainable raw materials (e.g. recycled materials) and efficient manufacturing techniques. However, there will always be some negative impacts with these processes.

Other parts of a new product's life cycle can also negatively impact the environment too, e.g. the transportation of a product causes the release of greenhouse gases.

Reusing a product several times is better for the environment than using a product once, before disposing and replacing it.

This is partly because reusing products rather than replacing them reduces the need to gather raw materials and manufacture new products. This reduces the negative environmental impacts linked to these processes, e.g. pollution.

In addition, much of a product's carbon footprint is generated during the early stages of its life cycle. The size of the carbon footprint generated when a product is being used, reused and disposed of is relatively small.

A product that is reused over a long period is therefore likely to have a smaller carbon footprint than repeatedly producing new products to replace products only used once. Reusing products therefore reduces greenhouse gas emissions and contributes less to global warming.

Reusing a product also diverts materials away from being disposed of in landfill sites. Landfill sites have a negative environmental impact, as they can pollute the surrounding land and water.

You should also include a conclusion in your answer that summarises your thoughts on the question.

Section C: Designing and Making Principles
Pages 170-178

20 How to grade your answer:

[No marks] There is no relevant information.

[1-2 marks] Brief consideration of the positive and/or negative aspects of the camera in relation to aesthetics/ ergonomics/suitability for users. Limited or no conclusion drawn. Points discussed may be limited to only positive OR negative aspects.

[3-4 marks] Positive AND negative aspects of the camera in relation to aesthetics/ergonomics/suitability for users are discussed and supported with sensible explanation. Conclusion drawn after considering positive and negative aspects.

20.1 Here are some points your answer may include:

Dark colour is stylish and appealing to the adult target market.

Large screen gives the camera a modern look and makes the camera look up-to-date to users.

The presence of many buttons and other controls give the camera a cluttered appearance, which may be off-putting to users.

Camera is bulky and rectangular in shape, which may be unattractive to users.

Different textures are used on some parts of the camera, e.g. on the sides of the rotary controls, compared to the rest of the camera, which looks attractive.

Simple, modern-looking symbols used to label the controls give the camera a stylish look.

Single, dark colour used, which some users may find boring and not very appealing.

20.2 Here are some points your answer may include:

Buttons and other controls are well-spaced and easy to use.

Some controls have a textured surface, which makes them easy to grip.

Shape of the camera doesn't appear comfortable to grip.

A strap is provided, which allows the camera to be carried hands-free when not in use, and within easy reaching distance.

The touch screen offers a simple way of using the camera as an alternative to buttons and other controls.

Controls are labelled in a contrasting colour to the camera's body, which makes them easier to read.

Some of the labels are quite small and close together, which may make them hard to distinguish for some users.

20.3 Here are some points your answer may include:

Water resistance allows the camera to be used in the rain/ damp conditions without being damaged.

Shock resistance means that the camera won't break if it is dropped onto a surface.

20 × zoom allows the user to focus and take a photo of something, even if they are a long way away from it.

HD video capabilities mean the user can record high quality video.

The GPS feature records where each photo was taken, which may be useful for users.

The screen could be easily scratched.

The screen is flat and may reflect some sunlight, so the user may find it difficult to use on bright days.

21.1 Any four from: e.g. the camera should have a raised grip *[1 mark]* to make it easier to hold *[1 mark]* / the camera should use large buttons *[1 mark]*, to make them easier to press *[1 mark]* / the camera should use big labels for the controls/a large text size for on-screen menus *[1 mark]* so they can be seen easily *[1 mark]* / the camera should use lightweight materials *[1 mark]* so that the camera is easy to pick up and hold *[1 mark]* / the camera must have controls/menus/settings that are easy to understand *[1 mark]* to make it simple to operate *[1 mark]*.

21.2 E.g. the large screen *[1 mark]* makes it easy to view detail shown on the screen. This is particularly important for elderly people who may be visually impaired *[1 mark]* / the strap provided *[1 mark]* allows the hands to be free to hold other items they may need (e.g. a walking stick) when the camera is not in use *[1 mark]*.

21.3 Explanation: e.g. a manufacturing specification is a series of written statements, or working drawings and sequence diagrams that explain exactly how a product will be made *[1 mark]*.

Examples: any two from, e.g. clear construction details explaining exactly how to make each part / the type/quantity of materials needed / the equipment required at each stage / the sizes/dimensions of each part / tolerances for each part / information on finishes / quality control instructions/what needs checking and when / costings of each part / health and safety precautions / tables/flowcharts showing the sequence in which tasks should be carried out / information about how long each stage will take *[2 marks]*.

22 Any two from: e.g. how efficiently the camera uses power *[1 mark]*. The lower the efficiency, the larger the power consumption and carbon footprint *[1 mark]* / The energy use of manufacturing processes *[1 mark]* as these contribute towards the carbon footprint of the camera *[1 mark]* / The types of materials selected for use *[1 mark]*. Sourcing materials can have environmental impacts, e.g. the production of metals and plastics can damage the environment during their extraction from the landscape *[1 mark]* / The sustainability of materials used *[1 mark]*. Sustainable materials (e.g. bioplastics and recycled materials) can have less of an environmental impact than non-renewable materials *[1 mark]*.

23.1 Metal: 0.2 × 360 = 72 people

Plastic: 0.55 × 360 = 198 people *[1 mark]*

You can also break the calculation down into a couple of simpler steps. For example, to calculate 55% of 360 people you can say: 100% of people is 360, so 1% of people is 360 ÷ 100 = 3.6. 55% of people is therefore 3.6 × 55 = 198 people.

23.2 Total number of people who said they would expect to pay £200 or less = 27 + 41 + 83 + 132 = 283

Percentage = (283 ÷ 360) × 100

= 78.6% (to 1 d.p) *[1 mark]*

23.3 How to grade your answer:

[No marks] There is no relevant information.

[1 mark] Answer contains a valid suggestion but no data is given to support it.

[2 marks] Answer contains a valid suggestion AND is supported by relevant data from figure 7.

Here are some points your answer may include:

Q1: e.g. the £176-200 range is the most popular, with 132 out of the 360 people expect to buy a camera for this price, so the camera should be designed so that it can be sold within this price range.

Q2: e.g. the camera should be able to store at least 801-1200 photos, as this was the most popular answer with 110 out of the 360 people that were questioned.

Q3: The main body of the camera should be made of plastic, as over half/55% of people questioned expressed a preference for plastic, with only 20% preferring metal.

There are lots of points you could have said for this question — you can get the marks as long as you have a valid reason that you have supported with data from the question.

24.1 E.g.

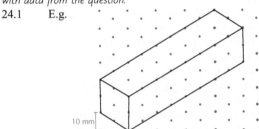

10 mm

Correct shape *[1 mark]*

Correct dimensions *[1 mark]*

24.2 The width of the box is too small *[1 mark]*, as it needs to be between 18.5 mm and 21.5 mm / has a tolerance of 20 ± 1.5 mm *[1 mark]*.

25.1 B = 40.00 ÷ 1000 = £0.04

C = 202.50 ÷ 2250 = £0.09 *[1 mark]*

25.2 E.g. length of thread needed for one pair of gloves in metres = 65 ÷ 100 = 0.65 m *[1 mark]*

Thread length needed to make 7000 pairs of gloves = 0.65 m × 7000 = 4550 m *[1 mark]*

Length of thread on a cone from supplier A is 1650 m, so the number of cones needed to give enough thread is 4550 ÷ 1650 = 2.76 which means that 3 cones are needed. Each cone costs £246.50, so the total cost of thread needed = 3 × £246.50 = £739.50 *[1 mark]*

Thread is only available in whole cones, so you'd need to buy 3 cones to have enough thread to make 7000 pairs of gloves. You can get all three marks for just stating the correct answer.

26.1 iterative design *[1 mark]*

26.2 Any two from: e.g. clients *[1 mark]* — this allows feedback to be gathered on whether the design/model/prototype fits with what the clients had in mind, so improvements can be made that suit the clients' wants/needs *[1 mark]*. / Users/target market *[1 mark]* — this allows feedback to be gathered on what potential users like/dislike about the design, so improvements can be made to make the product more appealing to them *[1 mark]*. / Experts/industry professionals *[1 mark]* — this allows the designer to benefit from their experience in the industry / allows the designer to get technical feedback about the design *[1 mark]*.

Glossary

alloy	A mixture of <u>two or more metals</u>, or a <u>metal</u> mixed with <u>one or more elements</u>.
anthropometrics	<u>Human body measurement</u> data.
automation	The use of <u>machines</u> to do a task <u>automatically</u> without much, or any, human input.
batch production	The <u>production method</u> used to make a <u>specific quantity</u> (a batch) of <u>identical</u> products.
CAD/CAM	Computer aided design/manufacture. <u>Designing</u> and <u>manufacturing</u> using a <u>computer</u>.
carbon footprint	The amount of <u>greenhouse gases</u> released into the atmosphere by <u>making</u>, <u>using</u> and eventually <u>reusing</u>, <u>recycling</u> or <u>disposing</u> of something at the end of its lifetime.
composite	A material made by <u>bonding</u> two or more different materials together.
continuous production	The <u>production method</u> used to make <u>large amounts</u> of a product <u>non-stop</u>.
corrosion	The <u>gradual destruction</u> of a material as it <u>reacts</u> with a substance, e.g. rusting of iron.
deforestation	Cutting down large areas of forest <u>without planting new trees</u> to replace the old ones.
design brief	The <u>instructions</u> that a <u>client</u> gives to a <u>designer</u> about what they want a product to <u>be like</u>.
design specification	A list of <u>criteria</u> that a product should meet.
ergonomic	A product that is <u>easy</u> and <u>comfortable</u> for people to <u>use</u>.
ferrous metal	A metal or alloy that <u>contains iron</u>.
fibre	A thin, <u>hair-like strand</u>. Fibres can be spun into yarn, or used as they are, to make fabrics.
filament	A <u>long continuous</u> length of <u>fibre</u>.
finite resource	A resource that will <u>run out eventually</u>, e.g. crude oil. Also called a <u>non-renewable</u> resource.
flexible manufacturing system (FMS)	A <u>set</u> of different <u>machines</u> which carry out the different <u>stages of production</u>. These <u>computer-controlled</u>, <u>automated systems</u> are designed to be <u>easy to adapt</u>.
hardwood	A type of wood that comes from <u>slow-growing trees</u> with <u>broad leaves</u> (mainly deciduous trees). It's usually <u>denser</u> and <u>harder</u> than softwoods.
integrated circuit (IC)	A <u>tiny</u>, <u>self-contained</u> circuit which can contain <u>billions of components</u>.
iterative design	A <u>design strategy</u> that involves constantly <u>evaluating</u> and <u>improving</u> a product's design.
knitted fabric	A fabric made from <u>yarns</u> held together by <u>interlocking loops</u>.
lean manufacturing	An approach to manufacturing that aims to <u>minimise</u> the <u>resources used</u> and <u>waste produced</u>.
manufactured board	A material made by <u>compressing</u> a mixture of <u>glue</u> and <u>processed pieces of wood</u> into <u>panels</u>.
manufacturing specification	A series of <u>written statements</u>, or <u>working drawings</u> and <u>sequence diagrams</u>, that tells the manufacturer exactly <u>how to make the product</u>.
market pull	When a product is made due to <u>consumer demand</u>.
market research	Asking the <u>target market</u> questions to find out their <u>likes/dislikes</u> (and so on) to help the designer understand what the target group <u>wants</u> from a product.
marking out	<u>Making a mark</u> in a material to show where it is to be <u>cut</u>, <u>drilled</u> etc.
mass production	The <u>production method</u> used to produce a <u>large number</u> of <u>identical</u> products on an <u>assembly line</u>.
microcontroller	A type of <u>integrated circuit</u> that can be <u>programmed</u>. Works as a <u>mini-computer</u>.
modelling	Making a <u>practice version</u> of a design or part of a design.

Glossary

modern material	A material that has been <u>developed</u> for a <u>specific application</u>. They are often developed through the <u>invention</u> of a <u>new process</u> or the <u>improvement</u> of an <u>existing one</u>.
natural fibre	A type of fibre that is harvested from <u>natural sources</u>, e.g. <u>plants</u> and <u>animals</u>.
non-ferrous metal	A metal or alloy that <u>doesn't contain iron</u>.
non-finite resource	A resource that can be <u>replaced</u> by <u>natural processes</u> as <u>fast</u> as it's <u>consumed</u> by humans, e.g. softwood trees in a plantation. Also called a <u>renewable</u> resource.
non-woven fabric	A fabric made from layers of <u>fibres</u> (not yarns) held together by <u>bonding</u> or <u>felting</u>.
one-off production	<u>Production method</u> used to produce a <u>single</u>, <u>unique</u> product at a time.
orthographic projection	A <u>2D scale drawing</u> of a 3D object showing the <u>front</u>, <u>plan</u> and <u>end views</u>.
planned obsolescence	When a product is <u>designed to become useless</u> quickly, e.g. a disposable razor.
printed circuit board (PCB)	A board with <u>copper tracks</u> that connect <u>components</u> in a circuit.
product analysis	<u>Examining</u> and <u>disassembling</u> a current product to get ideas for a new product or design.
production aid	A tool or technique used to <u>speed up</u>, <u>simplify</u> or help control the <u>accuracy</u> of a production process.
prototype	A <u>full-size</u>, <u>fully-functioning</u> product or system. It is built so that the <u>product</u> and <u>production methods</u> can be <u>evaluated</u> before the product is manufactured on a larger scale.
quality control	The <u>checks</u> that are carried out on a <u>sample</u> of materials, products or components throughout <u>production</u> to make sure <u>standards</u> are being <u>met</u>.
seasoning	The <u>drying</u> of wood to make it <u>stronger</u> and <u>less likely to rot</u> or <u>twist</u>.
smart material	A material that <u>changes its properties</u> in response to stimuli (a <u>change</u> in the <u>environment</u>).
softwood	A type of wood that comes from <u>fast-growing trees</u> with <u>needle-like leaves</u> (mainly evergreen trees). It's usually <u>less dense</u> and <u>softer</u> than hardwoods.
standard components	Common <u>fixings</u> and <u>parts</u> that manufacturers <u>buy</u> instead of manufacturing themselves.
staple fibre	A <u>short fibre</u>, e.g. cotton fibres are staple fibres.
stock forms	The different <u>shapes</u> that materials can be <u>bought in</u>.
sustainable	A sustainable process or material is one that can be used <u>without causing permanent damage</u> to the environment or <u>using up finite resources</u>.
synthetic fibre	A <u>man-made</u> fibre that is produced from <u>polymers</u>.
system	A collection of parts that work together to do a particular function. Made up of <u>inputs</u>, <u>processes</u> and <u>outputs</u>.
target market	The <u>group</u> of people you want to <u>sell</u> a product to.
technology push	When <u>advances in technology</u> drive the design of <u>new products</u> and the <u>redesign</u> of old ones.
thermoforming plastic	A type of plastic that can be <u>melted</u> and <u>remoulded</u> over and over again.
thermosetting plastic	A type of plastic that undergoes a <u>chemical change</u> when heated, which makes it <u>permanently hard</u> and <u>rigid</u>. Thermosetting plastics <u>can't be remoulded</u>.
tolerance	The <u>margin of error</u> allowed for a measurement of part of a product. Tolerances are usually given as an <u>upper</u> and <u>lower limit</u> e.g. 23 mm (± 2).
woven fabric	A fabric made by <u>interlacing two sets</u> of <u>yarn</u>.
yarn	A thread made by <u>twisting</u> fibres together. Yarns are <u>woven</u> or <u>knitted</u> to make fabrics.

Index

Index